40p

KU-589-831

No.

Date

poi

days
gran

o
unic

First published November 30th 1944
Second Edition 1950
Third Edition 1957
Reprinted 1959
Fourth Edition 1963
Fifth Edition 1966

5.1
CATALOGUE NO. (METHUEN) 02/4368/31

PRINTED AND BOUND IN GREAT BRITAIN BY
BUTLER AND TANNER LTD, FROME AND LONDON

A HISTORY OF
THE ROMAN WORLD
FROM 30 B.C. TO A.D. 138

BY

EDWARD T. SALMON
M.A., Ph.D., D.Litt., F.R.S.C., F.R.Hist.S.

MESSECAR PROFESSOR OF HISTORY AND PRINCIPAL OF UNIVERSITY
COLLEGE, McMASTER UNIVERSITY, HAMILTON, ONTARIO

LONDON: METHUEN & CO. LTD.
NEW YORK: BARNES & NOBLE INC.

METHUEN'S HISTORY OF THE GREEK AND ROMAN WORLD

DEMY 8VO.

METHUEN'S HISTORY OF THE
GREEK AND ROMAN WORLD

VI
A HISTORY OF THE ROMAN WORLD
FROM 30 B.C. TO A.D. 138

Naturally Ventilated Buildings

Naturally Ventilated Buildings

Buildings for the senses, economy and society

Edited by
D Clements-Croome

Construction, Management and Engineering,
University of Reading, UK

E & FN SPON
An Imprint of Chapman & Hall
London · Weinheim · New York · Tokyo · Melbourne · Madras

Published by E & FN Spon, an imprint of Chapman & Hall,
2–6 Boundary Row, London SE1 8HN, UK

Chapman & Hall, 2–6 Boundary Row, London SE1 8HN, UK

Chapman & Hall GmbH, Pappelallee 3, 69469 Weinheim, Germany

Chapman & Hall USA, 115 Fifth Avenue, New York, NY 10003, USA

Chapman & Hall Japan, ITP-Japan, Kyowa Building, 3F, 2-2-1 Hirakawacho, Chiyoda-ku, Tokyo 102, Japan

Chapman & Hall Australia, 102 Dodds Street, South Melbourne, Victoria 3205, Australia

Chapman & Hall India, R. Seshadri, 32 Second Main Road, CIT East, Madras 600 035, India

First edition 1997

© 1997 E & FN spon

Figure 7.1 © BRE Crown Copyright
All figures in chapter 9 © Ove Arup and Partners (except figure 9.X © Graham Gaunt)

Printed in Great Britain by TJ International Ltd, Padstow, UK

ISBN   0 419 21520 4

Contents

Foreword

Whether buildings are naturally or mechanically ventilated, they are designed and constructed to serve people and their requirements. An important requirement is that the indoor air quality should be felt as acceptable by most people and should have no adverse health effects. Furthermore, that the thermal environment is appropriate, i.e. felt neutral by most people with a minimum risk of draught.

These requirements are specified in different national and international guidelines and standards. For example, the CIBSE Guide provides guidelines on ventilation requirements and the thermal environment; EN ISO 7730 is a European and international standard on thermal comfort. A draft of a new European pre-standard, prENV 1752, has been elaborated on design criteria for the indoor environment in ventilated buildings. It covers both the thermal and the acoustic environment as well as indoor air quality. ASHRAE has also developed a draft of a new ventilation standard, 62R based on similar principles as the European draft.

The new standard European draft reflects an improved insight into human comfort requirements. An important new point in both the European and the American draft standard is that the building is acknowledged as a source of pollution as well as the occupants. In both documents there is strong encouragement to diminish this source, to decrease the ventilation requirement and save energy. This is a principle equally important for natural and mechanical ventilation.

Although some of the other comfort requirements were originally developed in mechanically ventilated spaces, they do provide general human requirements which apply whether the building is naturally or mechanically ventilated or air-conditioned. It has to be remembered that human expectations may be different and may influence human behaviour and the acceptability of the indoor environment as there has to be flexibility in using and composing standards.

Natural ventilation has been and will also be in the future, a dominant principle of ventilating buildings. Natural ventilation can indeed provide appropriate indoor environments during most of the year in cold or moderate climates. But methods of natural ventilation need to be developed further. Intelligent design of buildings and systems is as equally important for naturally ventilated buildings as for those ventilated mechanically.

It is essential that buildings can be adjusted to serve people It should not be the people who are required to adapt to the building. The building should be the servant, not the master. This sound ergonomic principle is the basis of building service engineering and applies whether the building is naturally or mechanically ventilated.

P. Ole Fanger

Professor D.Sc., Technical University of Denmark and President of the International Academy of Indoor Air Sciences

PREFACE

This book is based on a seminar held in July 1995 at The University of Reading on *Specifying Environmental Conditions for Naturally Ventilated Buildings*. It was organised by the Natural Ventilation Group of the Chartered Institution of Building Services Engineers. The views expressed in this book, however, are those of the individual authors and do not necessarily represent the corporate opinion of the CIBSE.

At a time when low energy building design is paramount the levels set for environmental conditions are crucial. Fresh air, daylighting, individual control and planned maintenance are vital features of a healthy building. It is important to ensure that an open view is taken as to whether a building is naturally ventilated or mechanically, or air-conditioned. All of these alternatives need consideration.

Knowledge about the performance of naturally ventilated buildings however, has remained comparatively scant, meanwhile standards for indoor climate have tended to emphasise active (mechanical) rather than passive (natural) airflow systems. There are many historical examples throughout the world of successful naturally ventilated buildings, but technology advancement has emphasised close control rather than the looser control offered by passive systems.

It is clear that decisions about natural ventilation and airconditioning need consideration of building orientation, form, layout, mass and fabric, and orientation. Also the client, architect, engineer and facilities manager have to work together as an interdisciplinary team. High standards of workmanship in building construction and careful selection of equipment are important.

It is hoped this book makes some contribution to knowledge about setting standards for naturally ventilated buildings.

I would like to acknowledge all the hard work of my fellow authors and to thank them for providing a rich fund of stimulating thoughts, ideas and applications. A special thanks to Karen Brown for laying out the text so carefully, and with originality; to John Jewell for producing the diagrams in the third chapter to his usual high standard.

Professor Derek Clements-Croome
Department of Construction Management & Engineering
The University of Reading

CHAPTER ONE

The Sense of Beauty – Role of Aesthetics in Environmental Science

Boon Lay Ong and Dean U Hawkes
The Martin Centre for Architecture and Urban Studies,
University of Cambridge, Cambridge, UK

Abstract

A model of natural beauty is proposed that relies on two recurring themes in many theories of aesthetics - *sensation* and *orderliness*. It is argued that the sense of beauty was perhaps evolved to help make quick and reliable judgements about the environment. This sense of beauty works in combination with our objective attention and the two forms of knowledge - *aesthetic* and *objective* - are complementary. Some of the evolutionary advantages that might accrue from having a sense of beauty are presented.

The aesthetic response is manifested emotionally and forms part of our experience of stress. Art helps us to deal with events in the real world by evoking emotions in us that imitate and thus mitigate real experiences. Architecture, though it serves a more functional purpose of shelter, also contributes to this mitigation of the environment and thus enhance our daily lives.

Environmental science can benefit from a more open-minded approach to the contribution of beauty and emotion to our perception of *heat, light* and *sound*. In allowing for the sense of beauty in environmental sensations, we must accept and take advantage of the fact that these sensations provide cues on how to react and respond to the world around us.

Keywords: aesthetics, beauty, emotions, environmental cues, objective knowledge.

Naturally Ventilated Buildings: Buildings for the senses, the economy and society. Edited by D. Clements-Croome. Published in 1997 by E & FN Spon. ISBN 0 419 21520 4

1 Introduction

Consider the following scenarios. A woman is shopping for a dress at a shop. She finally selects one and buys it. Upon going home, however, she finds herself disappointed with her choice and wonders how a sparkling new dress can become so ordinary in so short a time. A male office-worker buys a hamburger from a fast-food outlet. He takes a walk to a nearby park and finds himself a nice shady spot. He unwraps his burger, takes a bite and makes a face. Reluctantly, he eats enough of it to satisfy his hunger and throws the rest into the litter-bin. A couple of friends are having a holiday by the beach. The first day is drab and gray. It is not raining but they stay away from the beach, visit some shops and have a early night. The next day is bright and sunny. With a whoop of delight, they prepare for the beach and, *remembering the hazard of too much sun*, carefully pack their sun lotion and an alarm clock.

These scenes are fairly common. We know that clothes and jewellery appear more *beautiful* at the shops because of the carefully planned lighting levels and colour quality found there. Fast-food outlets are also carefully designed with bright colours, appropriate lighting, and high noise levels to encourage quick consumption of their food. Hamburgers from fast-food outlets sell better than those from the corner takeaway largely because the food is presented in an *aesthetically* more pleasing environment. Having a sun tan is a widely accepted *attractive* feature and proof of having had an expensive and exotic holiday overseas.

What these scenes have in common is that they are all examples of how our behaviour is determined by considerations of *beauty*. They also all demonstrate the importance of the *environment* in our perception of beauty. They show how we react to and are manipulated by aspects of the environment not generally thought to have emotive content - *heat, light* and *sound*. This ability to manipulate the thermal, lighting and acoustical environment to affect our behaviour is consciously applied by shopkeepers, airline pilots, fun-fair operators and fast-food retailers. Mothers know to turn down the light, put on some soothing music and make soothing sounds to help calm a baby and induce sleep. Indeed, manipulating the environment to provide the proper setting for any activity seems to be instinctively understood by almost everyone. Except, perhaps, environmental scientists.

In environmental science, the thermal, lighting and acoustical environment is regarded as a bland backdrop against which various activities are played out. The activity, or level of activity, which the human subject brings to the situation is thought to be determined by factors other than the environment itself. The subject is assumed to be unaware of the environment, or aware only at a subconscious level, until or unless the environment causes discomfort. Thus, the criterion for thermal comfort [10, 22] is based upon the notion of thermal balance: the right thermal conditions are those that balance the rate of heat loss by the human subject with the heat generated by the activity. In determining the operative temperature, the rate of generated heat is determined by the activity considered, which is statistically determined from laboratory tests and thought to be independent of the environment itself.

As a problem of physics, the thermal comfort analysis must be correct. The heat generated by the subject must be wholly balanced by the heat absorbed by the environment, or there will be thermal imbalance and excessive cooling or heating will result. Where thermal comfort fails is in its reflection of actual conditions in everyday life. As the examples at the start of this paper show, we respond to the environment by adapting our actions to it and we plan our day and activities by reading cues from the environment. What follows is an attempt to explain this behaviour and to look at some of the implications of expanding the assumptions of environmental science.

2 A model of natural beauty

As suggested above, we respond to the environment in an aesthetic manner. We go to and frequent places which we find beautiful. Commonly, beauty is thought to refer only to the visual experience. Although for the Western world at least the visual sense predominates over all others [19], beauty is not a purely visual experience. It can be argued that many of the pleasurable experiences related to *place* are not so much visual as they are *thermal*. Going to the beach, walking in the park, having a meal in a restaurant, reading a book in the conservatory, and having a party are all activities associated with certain thermal characteristics. It is difficult to imagine a place and think it beautiful under less than optimal thermal conditions - the beach on a cloudy day, the park on a foggy night, the conservatory on a dull English winter without heating.

It is proposed here that beauty is an instinct which was evolved to enable quick and generally effective evaluation of the environment. Our sense of beauty is a guide to all that is desirable to us. It stems from our accumulated experience of the world and is amenable to rational development and cultivation. There are two distinctive features to the presence of beauty - one, that it is determined by *sensations*; and two, that it is enhanced by an *orderliness* in its composition.

2.1 Sensation and knowledge

It is a truism to say that everything we know about the world arrives through our senses. In a very fundamental way, all knowledge is grounded on sense-data. However, our perception of the world is not of sense-data but of *objects* [2, 27]. The link between the sense-data and the objects is very intimate and we often confuse a *sensation* with the *object* that produces it. When we look at an apple, for example, we are not conscious of seeing colours and shapes first and concluding thereafter that the object is an apple. The identity of the object - *apple* - is evident immediately as part of the perception. We may argue that our consciousness of the *sensations* occurs later than our consciousness of the *object*.

Nevertheless, it is the sensations evoked by the object that stimulate our responses. We notice apples on a tree because of the colour of the ripe fruit. Conversely, we may not notice the abundance of fruits on the tree if they are unripe and green. The attraction we feel towards certain objects, like flowers, is linked to the *sensations* they produce in us. These sensations need not be colour alone, they could be smell (e.g. of roses), touch (fur), sound (waterfalls), or any of the other senses and could even be a combination of several senses. At the same time, the appeal of these sensations lies not just in the sensation alone but in its association with a desired object. Thus, the association of red with roses might be attractive while its association with blood might be repulsive.

It is this association of the sensation with the object, and the emotional attachment to certain distinguishing features, that is at the heart of aesthetic experience. The advantage of this aesthetic experience is its immediacy. Aesthetic experiences always have emotional consequences. When emotionally aroused, we respond quickly and often irrevocably. Research has shown that aesthetic responses can be made in less than 1/100th second and with incomplete information [23].

2.2 Order and knowledge

In just as fundamental a way as sense-data are related to knowledge, so too is order related to knowledge [2]. In terms of beauty, we generally see order as the pleasing arrangement of a given set of data or objects and do not normally associate that ordering to knowledge itself. Yet, the key to all knowledge lies in the arrangement of data into meaningful order.

At first glance, it is difficult to associate order with knowledge. We would normally associate knowledge to data and consider it to be a statement of fact while order is recognised as an artificial construct that we put onto the world. We might even talk of ordering knowledge as if we might gather knowledge from the world and then order it in a later separate operation. Deeper consideration will reveal, however, that we will not be able to even gather knowledge or data if we did not first order the environment into categories so that we know what falls within the data group we are gathering.

Even the most elementary form of life will need to be able to differentiate its environment to survive. At the very least, it needs to know what is threatening, safe, or consumable. Indeed, it is more important for this elementary form to know which category any passing object falls into than to determine what the object *really* is. Thus, we might imagine that the life-form need not differentiate between a seed and an ant if both are equally edible. On the other hand, it cannot afford to confuse two types of seeds if one of them is poisonous and the other is essential to its diet. Thus, objects are meaningful to us only in the way they fit *or don't fit* into the order we have created of the world for ourselves.

It may be argued that a sense for order *precedes* the acquisition of knowledge and that the more highly evolved an animal is, the more complex and profound is its sense of order [28].

2.3 Aesthetic and objective sense

The rational approach to knowledge best typified by science has been termed *objective* knowledge by Karl Popper [32]. There is some evidence to suggest that there is a complementary emotive sense of knowing that we may call *aesthetic* knowledge.

Objective knowledge is gained through attention. We pay attention to the evidence laid out, and decide for ourselves whether or not it may be believed. The learning of objective knowledge is a conscious act. However, research has shown [5] that even though background information may not register consciously, the presence of background or *contextual* information predisposes us favourably towards this information later when presented again as primary stimulus. Contextual information registers even if we are not fully aware of it. We are also aware that our disposition towards foreground information is influenced by the way it is presented - or by its context. Thus, contextual data affect us in two ways - they affect our disposition towards the object in the foreground, and they also affect our appreciation of objects in the background itself.

Research has also shown how objective knowledge may be converted into aesthetic knowledge [5]:

"initially, their views are unformed, so they think carefully about the issues involved. Then, as their attitudes develop, they cease such analysis, and have a more 'gut level' or emotional reaction (e.g., they become convinced they are right!). "

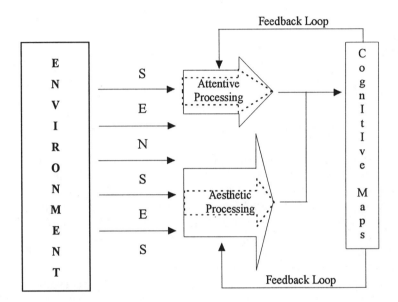

Fig. 1 Acquisition and organisation of aesthetic and objective knowledge.

The dotted lines represent the initial stages when the sense-data are not recognised. With repeated exposure, both objective and aesthetic awareness increase.

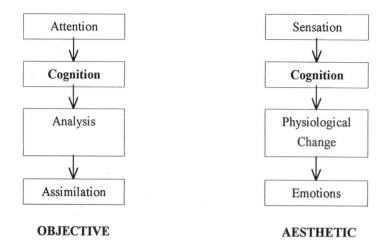

Fig. 2: Objective vs aesthetic processing

Fig. 1 shows not only that our sense-data are processed by both our objective and our aesthetics faculties, but that our ability to process the environment increases as our cognitive knowledge, represented by the cognitive maps we build, increases. Our aesthetic processing develops much more than do the objective faculty, which continues to deal with only one perception at a time. The aesthetic processing, on the other hand, increases more profoundly and we begin to see relationships, and features, and notice details that would have escaped us before or escape the untrained eye.

   Fig. 2 shows the different ways in which aesthetic and objective processing take place. While one leads to a greater understanding, the other prompts immediate and intense reactions. Where one serves to increase cognition, the other facilitates *re*-cognition. This predilection for easy *re*-cognition explains why our conception of beauty is attached to the twin factors of characteristic appearances and orderliness. Both of these factors facilitate easy recognition.

   The two modes are complementary and affect one another. This complementary role can be summarised as follows: while we use the *objective* process to help analyse and understand the world around us, the *aesthetic* sense is instrumental in developing our world-view and determines our response to the environment. The *aesthetic* sense is not infallible, as we know, and *objective* analysis is needed to complement its instinctive nature. However, the *aesthetic* sense is crucial in making correct judgements about how an object fits into our perceptual world. The sense of beauty is survival enhancing as it is concerned with an overall perspective and is swift in response. Perhaps arising from its survival enhancement value, our sense of beauty and the emotions it arouses are difficult to ignore and govern many of our actions - sometimes against our 'better' judgement.

3 Beauty in evolution

Our explanation of beauty is, so far, fairly simple. We are attracted to beauty through certain attractive features - smells, colours, shapes, textures, etc. - and by the presence of explanatory order. The presence of these two features are not accidental. Together and separately, they help to speed up our reaction to the world around us. The call to act is experienced in us as emotions and to understand beauty will require some level of understanding about emotions as well.

It is possible to argue that the sense of beauty exists even in the animal kingdom [7, 30]. We need not dwell too much upon this claim here but, if we allow the assumption that the sense of beauty exists even in the wild, it will be useful to see what evolutionary advantage, if any, is provided by this aesthetic sense. Some of the more obvious advantages are discussed below.

3.1 The need for vigilance

Life in the wild is fraught with danger. Predators try to take advantage of the times when their prey is least attentive. Since we are both predator and prey, we cannot afford to lower our vigilance while engaged in more pleasurable pursuits like eating or resting. Even as predator, we need to be on the lookout for traps and false signals. As a result, we must always reserve part of our attention to unforeseen eventualities. Our sense of beauty evolved from this need to be vigilant.

It is not always possible to visually locate an object directly. The food we crave for is not always easily available and must be hunted or found. This is achieved by looking for distinguishing features in the landscape that might indicate the presence of the desired object. Research has shown that even in the animal kingdom, responses to the environment are not direct [1, 6, 7, & 33]. Features like shape, smell and colour are more important in determining response than the presence of the object actually desired. For example, according to Cody[6], birds respond to general patterns like tree density and branch configuration rather than attempt to determine food availability directly. The ability to associate various features of the environment with respect to our needs is the source of our aesthetic sense.

3.2 Speed of response

It is important for our hunting-gathering ancestors, and even for ourselves today, to be able to make quick judgements of the environment. The inability to do so may mean death or lost opportunities. In the wild, this ability to make spontaneous decisions is perhaps even more crucial, often literally a case of life and death.

Typically, an aesthetic reaction is very quick and very definite. For example, Kaplan and Kaplan [23] in their investigation on landscape preferences, found that their subjects were able to make a choice of sites from black and white photographs flashed on screen for less than *one-hundredth of a second!* Subsequent visits to the actual sites and longer exposures to the photographs made no difference in terms of their liking or disliking the landscape, only to the degree of preference.

The key feature in being able to make such quick decisions lies in the ability to make judgements based on significant details in appearances - extended here to include all sensations. Thus, we can tell that a fruit is ripe from its colour and smell. We conclude that a water body is safe to drink because we see live animals drinking from it, and we move upstream, not downstream, to ensure a cleaner supply. Dark clouds portend rain and we take note of where shelter may be found nearby. This dependence on the senses is not accidental. It has been suggested [5, 26] that our senses evolved primarily to allow us to decipher the environment at distances, and from weak stimuli.

This ability to make quick, and usually reliable, decisions is of great advantage. It helps to procure food that can run away, escape enemies that are larger and perhaps faster than ourselves, and claim territory ahead of competitors.

3.3 Discerning details

We cannot afford to take into account all aspects of the appearance of the object. There is a separation of important features from those that are incidental. There are also general rules that are useful: most fruits ripen to shades of yellow, orange, brown or red; the sky is blue when clear, other colours in the sky suggest impending change (dawn, dusk, gathering storm). These rules become so ingrained that the features themselves evoke an immediate response.

But appearances deceive. So the ability to discern detail and read appearances correctly is very valuable. Lynch [25] speaks of the ability to navigate by the stars in some 'primitive' societies:

"The ability to distinguish and orient in these resistant environments is not achieved without cost. The knowledge was usually limited to specialists. Rasmussen's informants who drew his maps were chiefs - many other Eskimos could not do it. Cornetz remarks that there were only a dozen good guides in all south Tunisia. The navigators in Polynesia were the ruling caste. "

It is reasonable to suppose that if this ability to read the environment is important, then there might be some evidence of it in the animal kingdom. For example, we might expect the leader of a pack of animals to lead in navigation - a task requiring an ability to read the environment well. Hans Kummer in his study on *Primate Societies* [24] found that this was not the case. Surprisingly, travel can be initiated by any member of the tribe. In male dominated species, the lead may even be taken by the females. Often, primate groups are led jointly by several adults. Although this meant that the routes taken are a result of compromise, there is no sign of aggressive conflict between the members. He found only one example, the mountain gorilla, of the kind of navigating leadership that we have supposed to operate. Yet, perhaps the evidence does not contradict our hypothesis.

Take the following example of a small band of hamadryas baboons led by two males, Circum and Pater, setting off to find new feeding grounds [ref. 24, pp 64-65]:

"During most of the day, Circum tried to lead the band to the north, where the rest of the troop had gone a few days before. He was adamantly opposed by the older and more influential Pater, who insisted on a trip to the southwest. As a result, the trip assumed a peculiar zigzag pattern, but the two males never showed any signs of impatience or aggression, not did the two groups separate. At 2 pm, Circum finally abandoned his northward trend and preceded the band in the direction that seemed to correspond with the intentions of the older male."

Perhaps the importance of navigation is such that dominance in the pack, which is usually determined by *strength*, is surrendered to the more *astute* animal. It is perhaps significant that the hamadryas band finally followed the direction suggested by the *older* monkey but still with the younger in the lead. Perhaps that is also why other *adult* members of the tribe can take the lead when travelling. The same reasoning will also explain why such decisions were made without aggressive behaviour.

3.4 Recognition and communication
Just as it is important to be able to read the environment correctly, it is just as important to be able to transmit signals correctly. Language is a distinguishing characteristic of human societies and the suggestion that beauty, as expressed in art and language, is a form of communication might be seen as obvious, perhaps even self-evident. The notion that art is just visual or sensory stimulation which does not signify anything is very easily disproved by the perusal of virtually any contemporary text of art [e.g. ref. 11]. However, there is also a wide range of literature on the *attractive* devices used by animals to tell other animals, of their kind as well as those who are not, when they are angry, scared, healthy or reproductive [e.g. refs. 1, 8, & 33]. Many of these devices of beauty - the antlers of the stag, the tail of the peacock, the colourful buttocks of some monkeys - endow their possessors with a burden as well. The possession of beautiful features often make the animal easier to notice and physically slow it down.

Perhaps the most amazing example of communication in animals may be found in the honey bee [1]. The dance that honey bees use to communicate directions to their fellow worker bees is complex not only in their ability to describe accurately to other bees geographical data but also in that the information is referenced against the position of the sun. Amazingly, the recipient bee is able to correct for the movement of the sun and still find its way to the source.

The communicative dance of the honey bee clearly holds a great deal of information and its relationship to beauty, except in the eyes of the amazed human beholder, is perhaps hard to see. This is so only because the modern conception of beauty excludes the function of communication. Within our conception, however, beauty is *primarily* concerned with communication and what the honey bee example shows is the relationship between the perceived environment and its communication *via beauty* - appearance and order. The role of the honey bee dance is similar to the traditional functions of art and literature in many human cultures.

3.5 A shared language
The role of beauty in recognition and as communication is so fundamental as to favour a common language. What is beautiful to one animal is likely to be so to other animals and the interpretation of particular beautiful features is likely to be understood by different species as well. Animals which do not conform to the norm will have a lower chance of survival and in the process of elimination, there will be a tendency for the accumulation of commonly understood signals. It is not necessary to survival for all animals to understand equally all signals from the environment or from all animals. Each animal will have its own sphere of existence and only those signals that are pertinent within each sphere need to be correctly interpreted. Correct interpretation will depend also on the animal since what is predator to one may be prey to another, what is food to one may be poison to another, and what repels one may be ideal hideaway for another. Perhaps paradoxically, it is the very ability of an animal to realise what might be poisonous to most of the competing life-forms and take advantage of it that suggests most strongly that the aesthetic language must be shared.

There is of advantage to the ecosystem and to the animals individually to share this aesthetic understanding.

That such cross-understanding exists is fairly well documented. As we have suggested earlier, it is just this commonality of understanding which allows each animal to carve out its own niche. Unless the animal can tell how other animals read the environment, it will not be able to determine the niche most advantageous to itself. Darwin, in one of his less well-known but still highly significant work *"The expression of the emotions in man and animals"* [8], was able to show that the ability to recognise facial expression is not only innate but also universal. Other researchers have since developed the work further and we are able now to even postulate how certain facial expressions, like smiles, might have evolved [26].

4 Sense and emotions

Aesthetic responses are so strong and immediate that our bodies usually react before our minds can take conscious control of our actions. Sometimes, even when we are conscious of the illusory nature of the situation - in the cinema, for example - we may still be unable to control the emotions aroused.

Emotions evoked by transient and artificial situations like film or art can usually be subdued soon after the event. Real-life events, on the other hand, though often less extreme are more enduring. Emotional events of either sort cause stress.

4.1 Stress and environmental science

Stress is a result of our body maintaining homeostasis. A more detailed account of our perception of thermal stress and its consequences for environmental science is presented in a later chapter in this book (see *From Homogeneity to Heterogeneity*). For the moment, what we want to highlight is the aesthetic implications of environmental stress.

We respond to the environment aesthetically. We find what is desirable in the environment beautiful and what is not ugly. The association of desirability with beauty extends across all our senses. Thus, artists evoke our senses through the depiction of scenes and/or the use of media. The senses that art arouses are not just visual but also thermal, giving us a sense of warmth or coolness; or tactile, a sense of roughness and surface; or olfactory, a sense of putrid decay or aromatic sensuality; and so on. We normally associate art with the visual, but this association is a consequence of our current primacy of vision and of the media most artists employ. But it is just as possible to have art and aesthetics in other media. The master chef pleases us through our taste buds as well as visually and we enjoy and recognise the art of cuisine as a valid form of aesthetic experience. The use of perfume is another example of recognised aesthetic experience.

If we acknowledge that this aesthetic response to the environment permeates through all our behaviour, then it follows that environmental science needs to recognise this regulation of physiological responses by our emotions, and ultimately by our sense of beauty, and seek an approach that reflects this reality.

4.2 Art as mitigating experience

We have suggested earlier that the emotions evoked through art are less stressful than real events. In terms of stress, what art does is to raise the pertinent thresholds[29]. This happens both because we engage with art in the abstract, which prepares us for real-life events, *and* because art allows greater, yet safer, sensory and emotional experiences.

While art arouses the senses, the medium through which this occurs is recognisably artificial. Thus, while we can be saddened by the tragic event in art, we are not as saddened as we might be by the loss of a loved one, even a pet. On the contrary, we react to the emotions evoked by art mentally, if not intellectually. Through the provocation of art, we are able to contemplate perhaps dispassionately emotions which if they occur in our daily life, we would not be able to handle quite so successfully. In doing so, the range of our everyday experience is expanded and enriched. In a way not unlike the training that athletes undergo before they compete, we are better prepared to face the unexpected in real life because we have been primed in a more controlled artificial milieu.

The environments formed by good architecture often have this mitigating quality of art. The glory of nature is presented without exposing the user to its dangers. Atria & conservatories are popular for this reason: they allow a visual connection with nature while protecting us thermally.

It is of the essence to architecture, of course, to provide shelter and protection. But the functions of architecture go beyond that of merely keeping out the elements. There are degrees to which we will accept being isolated from the rest of the world. Keep it out too much and we feel imprisoned. Keep it out too little and we will not stay. Good places are almost synonymous with sanctuary and repose. The mitigating effect of architecture may be less evident to the reader than that of art. One way to bring out this role of architecture is to consider its effect on our appreciation of art. The impact of art is strongly influenced by the architecture which surrounds it. The same artwork placed in a museum will evoke different emotions if it is later seen in a 17th century house, or encountered in a warehouse. In the same way as architecture can affect our perception of the art it contains, it also affects our perception of the environment and events that it frames.

The aesthetic experience afforded by art and architecture is important because it expands and educates our sense of beauty. In doing so, we have a heightened appreciation of the world around us and a greater awareness of the possibilities that exist for us.

5 Design implications

Today, we have a different, if not better, understanding of how buildings work than before. New technologies exist and new ways of living in buildings are being explored. This difference in understanding must acquire new expressions of form.

There is an environmental *design window* appropriate for any particular activity. This window sets certain parameters, like rules of a language, to how the built environment can support various activities or events. These parameters involve the way the environment is received by us aesthetically through our senses - *heat, light & sound*.

Existing thermal standards tend to work towards thermal neutrality. Such criteria are perhaps too narrow and too simplistic to reflect the attraction of real *places*. If we accept that thermal sensation provides aesthetic information, then the *design window* may be wider, and more complex, than we have supposed. Beyond the provisions of comfort, the aesthetics of the thermal environment would involve the provision of certain contextual cues.

5.1 Cues about activity

The *aesthetic reading* of the environment means that sense-data - even heat, light & sound - cannot be viewed neutrally. Warm environments tend to tell our bodies to slow down and generate less heat. Cool environments, on the other hand, tend to stimulate activity. We not only respond physiologically to heat and cold by sweating or shivering but we also seek suitable places for particular activities and modify our active-ness to suit the thermal ambience. We have even developed cultural and social habits that respond to the thermal environment [18].

Nick Baker [3, 4] has suggested that there is a *cognitive tolerance* in our response to thermal stress. He pointed out that thermal stresses related to desirable activities like swimming are endurable because the stresses are recognised as integral to the enjoyment of the activity.

5.2 Cues about operation

In the tropics, we tend to avoid the sun and seek shade. The reverse happens in Western countries. When we use a fireplace, we increase the size of the fire when we are cold and move closer to it. When we go to sleep, we put off the fire and allow the room to cool down as the bed warms up. We watch the glowing embers so that our faces may gain from the fire's dying radiation.

Modern designs for heating and cooling lack these self-evident cues about operation. Dr Hawkes began work on this almost 20 years ago in the Martin Centre [12, 13]. Much more remains to be done.

5.3 Cues about place

Rev. Michael Humphreys [20, 21] have long argued that thermal criteria ought to relate to the places to which they are applied. It is far better to work with traditional adaptations than to work against it. The imposition of an unchanging thermal standard has meant that, when we inhabit these spaces, we are not thermally aware of the external environment.

5.4 Cues about time

The desire for windows[34], and for some clues about the prevalent weather outside, is fairly well established in environmental literature. Such a desire also has thermal implications. Windows not only provide visual relief and visual connection but also tells us about the thermal conditions outside. The temporal occurrence of these thermal conditions - whether it is raining in the middle of the afternoon, or at night - affects our response. A slight breeze is likely to be more welcome in the morning when we are more active and there is a greater probability of later sunshine than in the evening.

6 The need for greater sophistication

As our understanding of environmental needs grow, it would seem that the criteria for good design will grow unmanageably complicated. This is so only if we continue to design every project from scratch. As Hawkes [14] has described it, every existing building is

"a *store* of accumulated experience which contains all previous solutions and which will be enlarged in the future with the addition of new examples inspired by changing building technology, organisational ideas and physical, social and cultural environments." pg 481

Environmental science must be deemed to have reached the point of development when the complexity of the subject can be addressed in its fullest [29]. In doing so, there is a need to marry the precision of laboratory findings with the complexities of everyday life. We need to develop typologically specific studies which will embody all that we know about the specific building type and extend the analysis to incorporate findings from historical, psychological, social and other fields not presently consulted. The availability of sophisticated design aids will allow us to study theoretically, at least, the parametric limits of various design types. Such comprehensive studies will provide much needed theoretical underpinnings to future design.

6.1 Presence of nature (plants)

Plants are often employed for aesthetic reasons. However, research has shown that plants are not only psychologically beneficial [23], they also modify the micro-climate in positive ways. For a start, the pleasant colours and reflectivity of plants tend to reduce glare. They are, of course, responsible for most of the oxygen in the air. Plants, and microbes in the soil, help to clear the air of dust & pollution. Current work at the Martin Centre shows that plants keep the microclimate at an even keel - reducing high temperatures and keeping the air moist [30].

The use of plants in architecture is one of the least tapped and most promising possibilities in the development of an ecologically sound architecture. Throughout our evolutionary history, we have depended on plants for many of our basic needs - food, energy, clothing, shelter, raw material. We now know that they are largely responsible for the life-supporting climate on earth. The evolution of the animal kingdom is a story of the complementary evolution of plants as well. This principle of co-evolution must suggest that our future development depends on our ability to ensure the corresponding success of plants - and of the rest of the earth's ecosystem - as well.

7 Conclusion

Taking on board our aesthetic sensibilities in the development of environmental science does not mean that we must now embrace traditional superstitions that have emotional appeal but no scientific basis. On the contrary, the two forms of knowledge - *objective* and *aesthetic* - are complementary and efforts must be made by scientists to synthesise the two. This we can only do if we consider *emotions* as an additional variable in our criteria for thermal environments.

In conclusion, we would like to reproduce a quotation from a favourite scientist. His words were written almost 60 years ago and are presented here out of context. We have also taken the liberty to change one word - from *thinking* to *living*. The spirit reflected in this quotation sums up well our central thrust:

"The whole of science is nothing more than a refinement of everyday thinking. It is for this reason that the critical thinking of the physicist cannot possibly be restricted to the examination of the concepts of his own specific field. He cannot proceed without considering critically a much more difficult problem, the problem of analyzing the nature of everyday *[living]*. "

Einstein, Albert (ref. 7, pg 290)

8 Acknowledgments

An earlier version of this paper was presented at the CIBSE National Conference 1995 held in Eastbourne 1-3 October 1995. The present paper has been revised to tie in better with another paper, *From Homogeneity to Heterogeneity*, which is published in this book and which repeats some of the material presented here. Nevertheless, it has not been possible to avoid some repetition without substantially destroying the integrity of the present paper.

9 References

1. Attenborough, D. (1990) *The Trials of Life: A Natural History of Animal Behaviour,* Collins/BBC Books, London.
2. Ayer, A.J. (1956) *The Problem of Knowledge* Pelican Books, Middlesex.
3. Baker, N. V. (1994) *The Irritable Occupant* The Martin Centre Research Society Lectures - unpublished.
4. Baker, N. V. (1995) *Adaptive Opportunity as a Comfort Parameter* Workplace Comfort Forum, 22-23 March, RIBA, London.
5. Baron, R. A. (1989) *Psychology: The Essential Science* Allyn and Bacon, Boston, USA.
6. Cody, M. L. ed., (1985) *Habitat Selection in Birds* Academic Press, New York.
7. Darwin, C. (1859) *The Origin of the Species* Murray, London.
8. Darwin, C. (1965) *The expression of the emotions in man and animals,* University of Chicago Press, Chicago, Illinois.
9. Einstein, A. (1954) *Ideas and Opinions* Carl Seelig (ed.), A Condor Book, Souvenir Press, New York.
10. Fanger, P. O. (1970) *Thermal Comfort* Technical Press, Copenhagen.
11. Gombrich, E. H. (1995) *The Story of Art* (16th ed.) Phaidon Press, London.
12. Hawkes, D. U. & Willey, H. (1977) User response in the environmental control system *Trans. Martin Centre Arch. & Urban Studies,* Vol 2, pp 111-135.
13. Hawkes, D. U. & Haigh, D. (1980) *User Response in Environmental Control* Martin Centre Technical Report, Martin Centre, Cambridge.
14. Hawkes, D. U. (1976) Types, norms and habit in environmental design *The Architecture of Form* by March, L., ed., Cambridge University Press.
15. Hawkes, D. U. (1988), Environment at the Threshold *A New Frontier: Environments for Innovation* (Proceedings: International Symposium on Advanced Comfort Systems for the Work Environment) by Knoner, Walter M (ed) May 1988 pp 109-117, Center for Architectural Research, Rensselaer Polytechnic Institute, Troy, New York - to be reprinted in Hawkes, D. U., 1995, "The Environmental Tradition: Studies in the Architecture of Environment", E. & F. N. Spon, London.
16. Hawkes, D. U. (1992) The Language Barrier *AJ,* Vol 5, pp 22-25.
17. Hawkes, D. U. (1995) *The Centre and the Periphery* The Martin Centre Research Society Lectures - to be published in *Architectural Research,* Vol. 1, No. 1.
18. Heschong, L. (1979) *Thermal Delight in Architecture* MIT Press, Cambridge, Massachusetts.
19. Howes, D. ed., (1991) *The Varieties of Sensory Experience* University of Toronto Press, Toronto & London.
20. Humphreys, M. A. (1975) *Field Studies of Thermal Comfort Compared and Applied* Building Research Establishment Current Paper CP 76/75, August.
21. Humphreys, Rev. M. A. (1995) What causes thermal discomfort? *Workplace Comfort Forum* 22-23 March, RIBA, London.
22. ISO 7730, *Moderate thermal environments - Determination of the PMV and PPD indices and specification of the conditions for thermal comfort* First edition, 1984-08-15, UDC 331.043.6, Ref No ISO 7730-1984 (E).

23. Kaplan, R. and Kaplan, S. (1989) *The Experience of Nature,* Cambridge University Press, Cambridge, UK.
24. Kummer, H. (1971) *Primate Societies: Group Techniques of Ecological Adaptation* Aldine Atherton, Chicago.
25. Lynch, K. (1960) *The Image of the City* MIT Press, Cambridge, Massachusetts.
26. McNaughton, N. (1989) *Biology and Emotion* Cambridge University Press, Cambridge, UK.
27. Merleau-Ponty, M. (1962) *The Phenomenology of Perception* Routledge and Kegan, London.
28. Morris, D. (1967) *The Biology of Art* Aldine Atherton, Chicago.
29. Ong, B. L. (1995a) *From Homogeneity to Heterogeneity* - this book.
30. Ong, B. L. (1995b) *The Biology of Beauty* Martin Centre Research Society Lectures - unpublished.
31. Ong, B. L. (1995c) *Place and Plants in Architecture* Ph.D. Thesis, Cambridge University - unpublished.
32. Popper, K. (1979) *Objective Knowledge - An Evolutionary Approach* rev. ed., Clarendon Press, Oxford.
33. Tinbergen, N. (1951) *The Study of Instinct* Oxford University Press, Oxford.
34. Ulrich, R. S. (1984) View from a window may influence recovery from surgery *Science,* 224, pp 420-421.

CHAPTER TWO

From Homogeneity to Heterogenity

Boon Lay Ong

The Martin Centre for Architecture and Urban Studies,
University of Cambridge, Cambridge, UK

Abstract

It is generally accepted that the environment imposes stress on people. However, the relationship between stress and the environmental response scales in common use is yet to be explicitly made.

This paper presents an explanation of stress that directly relates psychological stress to environmental responses. This is achieved by defining thresholds of stress that parallel the criteria used to determine environmental response. Using this, the function of art and architecture can be described in terms of the reduction of stress through the mitigation of experience. This mitigation occurs because art depicts reality (but is not reality itself) and through its artificial framework allows a wider range of experiences than is otherwise possible.

It is argued that this mitigation of experience supports the design of a *heterogeneous* rather than a *homogeneous* environment. Such heterogeneous environments succeed not only because artificial, or built, environments are inherently *safer* than the natural but also because they provide positive stimulation which promotes stress resiliency. Not least, they also offer more choice to the user.

Keywords: ISO 7730, stress, sensorium, heterogeneity, environmental criteria

1 Introduction

In August last year(Ong 1994), I presented a paper which used the ISO 7730 to demonstrate the inadequacy of single temperature standards and suggested that the solution is to have a *double* standard - one for the *ambient* environment and one for the *local*. This suggestion of a double standard may be seen as a variation on the theme of

Naturally Ventilated Buildings: Buildings for the senses, the economy and society. Edited by D. Clements-Croome. Published in 1997 by E & FN Spon. ISBN 0 419 21520 4

environmental diversity that has been proposed elsewhere (Hawkes 1988).

I should like to now develop the theme further: first, to reinforce the importance of environmental diversity; second, to provide an account of environmental stress; and finally, to suggest how a more sophisticated approach to environmental science may be developed.

2 Re-interpreting the ISO 7730

The paper (Ong 1994) for the IATS 94 conference showed that there is a range of acceptable temperatures for any given activity if we take into consideration individual variations in *size* and *body weight* (see Fig.1). This range of acceptable temperatures changes with increasing activity and suggests that we can tolerate more changes in the environment if we are active than if we are sedentary.

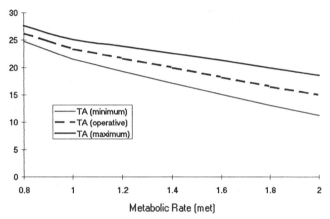

Fig. 1: Range of operative temperatures vs activity (measured in mets) - Ong 1994

The operative temperatures in Fig.1 were derived from the consideration that our metabolic rates vary with body weight and size. This consideration is supported by current wisdom in the fields of diet and exercise physiology (McArdle, et al., 1991). This variation of metabolic rates is based only on the variation of the rate of energy expended proportional to the mass moved and does not take into account the fitness, sex or age of the individual, which will affect the rate and efficiency at which the mass is moved and therefore the amount of energy spent.

It can also be shown that for different PMV indices, the operative temperatures also change with increasing activity. Fig.2 shows this variation in operative temperatures for the various PMV indices. As expected, the neutral temperature (PMV=0) drops as activity increases. While the general trend of operative temperatures is to drop as activity increases, the range of temperatures within each PMV index increases with increasing activity.

Similar charts can be generated if we vary other parameters in the ISO equation - clothing level, air vs radiant temperatures, wind speed, etc. If we further relate metabolic rates to physiological factors like age, sex, and fitness, then we may generate yet more charts showing, again, a wide range of operative temperatures.

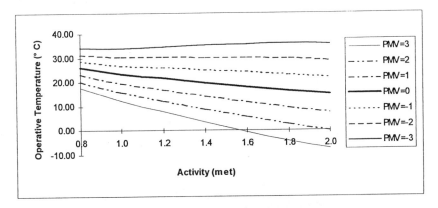

Fig. 2:   Variation of operative temperature with changes in activity showing a) the reduction of temperature with increasing activity, and b) the increase in tolerance with increase in activity. Based on ISO 7730.

The point of this exercise is to demonstrate that the variation in individual needs, even within the scope delineated by the ISO, is sufficient to suggest a great discrepancy between predicted comfort levels and actual conditions. A recent study (Newsham & Tiller, 1995) observed that *"around 90% of the variance in thermal sensation/preference vote was unexplained by indoor air temperature and other measured variables."*

To understand further this deviation between statistically derived standards and observed behaviour, we need to look into the mechanics of environmental stress.

3 Stress and the Environment

Stress is a result of the body attempting to maintain homeostasis despite changing external conditions. Discovered by Hans Selye (1973), and developing from the work of W D Cannon (1932), stress is now so widely accepted as to have entered public consciousness.

Despite the general acceptance that stress comes from the environment, the actual relationship between the concerns of environmental science - heat, light and sound - and stress has not been explicitly made (see Evans, 1982, for example, for a more detailed general discussion of environmental stress). The following explanation shows how this relationship might exist.

At very low levels of stress (Fig.3), our bodies cope with an ease that is largely unconscious. Depending on the nature of the stressor - from infection/digestion to threat/pleasure - the appropriate defense mechanisms are instigated and the stressor dealt with. At this low level, a variety of hormonal and immune responses are possible and it would be tediously long to describe them all.

BL Ong

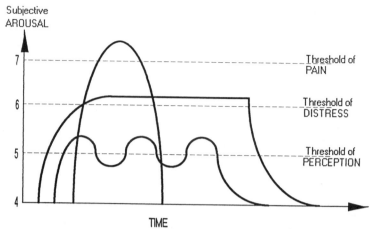

Fig.3: A diagram showing the various thresholds of intermediate stress. The bold lines indicate different stressors. It should be noted that with the passage of time, the thresholds will change - becoming less conscious as we *adapt* to the situation or becoming more significant if we fail to adapt.

However, the effects of stress can accumulate if the same stressor is present for a long time or if the intensity of the stressor is increased, or if other stressors present themselves. Thus, our arousal by any stressful event depends on our level of stress at that particular point in time. A level of stress, which we shall call the *threshold of perception*, is reached where we become conscious of the presence of the stressor. We experience a sense of discomfort or disequilibrium. If our bodies are allowed to adapt to this level of stress, and a conscious acceptance of discomfort is required for this, then the stressor is again dealt with adequately and a new equilibrium is attained. This ability to maintain different levels of equilibrium is called *heterostasis* (Selye 1973). If the stress continues, either in time alone or in intensity as well, a point is reached where we will actively seek ways to remove or reduce the stress. This is the *threshold of distress*. From this point on, the stressor commands our attention. Between this level and the next, the body is under extreme strain. It is between these two levels that professional athletes and weightlifters train. At this stage, our discomfort with the stressor extends beyond the bounds of the event itself. For very stressful activities, like weightlifting, for example, the recuperative period can extend to two or more days. The upper limit of stress may be called the *threshold of pain*. Beyond this limit, permanent tissue damage and/or disease results.

The significance of these thresholds is that they are subjectively perceptible and recognisably distinct. As there can be too *little* as well as too much stimulation, the three thresholds identified in Fig.3 are mirrored below the line of neutrality. The full picture incorporating both the over-stimulation and under-stimulation thresholds parallel the 7-point rating scale in common use. The ratings of thermal standards like the ASHRAE or ISO scales (Table 1) may therefore be regarded as an application of these stress thresholds.

Bedford		ASHRAE		
Much too warm	1	Cold	1	+3
Too warm	2	Cool	2	+2
Comfortably warm	3	Slightly cool	3	+1
Comfortable	4	Neutral	4	0
Comfortably cool	5	Slightly warm	5	-1
Too cool	6	Warm	6	-2
Much too cool	7	Hot	7	-3

Table 1: Comfort and thermal sensation scales in common use. Note the reverse thermal relationship between the Bedford and ASHRAE scales. (Cooper 1982)

Notice that the deviation from neutrality exists on both sides - with too much as well as too little stimulation. Consequently, there is a *window* of environmental conditions which is appropriate for any particular behaviour. This *window* is the *design window* within which environmental design can operate. For attentive work, the *design window* is narrow, defined within the *thresholds of perception*. For non-attentive work, like resting or having a conversation, the *design window* is wider, as a slight discomfort is more easily accommodated. As well, the pleasure of enjoying the scenery, for example, may be a significant compensation. Finally, as argued earlier, an increase in metabolic rates will also increase the width of the *design window* (Fig. 2). This increase in metabolic rate can be brought about by both physical activity and psychological arousal.

4 Detailed aspects of stress

It must be emphasised that stress is seen as the body's *general* response to changes in the environment. The effects of *different* stressors are accumulative and even desirable stresses, those that we enjoy like dancing and play, can be detrimental if taken to extremes. The symptoms of stress are general and do not reflect the stressor. In thermal stress, for example, stress symptoms are less those changes associated with heat or cold (sweating, shivering, etc) but more the accumulated effects on the body (tiredness, susceptibility to illness, etc.).

The effects of stress can be modified by several factors, including *diet*, *attitudes*, *rest* and *exercise* (see Glass 1977, for example). For the purpose of this paper, the influence of diet may be discounted.

4.1 Stress & immunity to disease

We accept that, under distress, we are more susceptible to falling sick. There are several reports on stress related susceptibility to illness (e.g. Jokl 1977, and Borysenko 1984). This is true not only with physically stressful activities like sports and athletics, but also with psychologically stressful activities like office work - or traumatic experiences. On the other hand, moderately stressful and pleasurable activities - like going on a holiday, or exercise - are generally regarded as healthy. Some studies have shown that positive emotional states are correlated with increased immunity (reported in Thayer 1989, pg 37-38; see also Cousins 1979). One such correlation was found after viewing humorous films or videotapes, which suggests one possible and useful function for humour.

4.2 Psychological receptiveness to stress

The various thresholds of stress are shifted according to how the stressor is perceived psychologically. If the stressor is part of a desired event, then the thresholds are effectively raised. On the other hand, if the event is dreaded, then the thresholds are lowered. This may be explained in terms of the mind anticipating the outcome of the impending event and initiating the appropriate response accordingly and pre-emptively. This internal conflict between engaging with the stressor and avoiding it may be regarded as the psychological component of stress.

This helps to explain the significance of coping strategies in dealing with stressful events. The psychological reassurances of friends, relatives and professionals do not remove the stressor itself but remove the internal conflict that constitutes the psychological aspect of stress. The wisdom of the body, to use W B Cannon's (1932) phrase, is allowed to operate and the thresholds are lifted. The lack of such social support may be used to explain why some patients succumb to the toils of disease. Instead of spurring the body to fight back, in these cases, the patients convince themselves that the fight is useless and the body succumbs.

4.3 Warming up & cooling down

We normally allow ourselves some time to warm up and cool down before engaging in any exercise. These preparatory and recuperative periods are important to all events of great stress. Psychological events that are unexpected are more shocking than events that can be anticipated. Also, important changes in our lives, like moving into a new house or getting married, require that we spend some time adjusting to the new conditions both before and after the event.

It is important in architectural design to provide for the mediation between different activities.   Intermediate spaces like foyers, cloakrooms and anterooms play an important role in helping us make the transition from the outside to the inside, and between different spaces. Spaces which combine the inside and the outside, like verandahs, balconies, bay windows and conservatories, are often regarded with particular affection. Visually, the appearance and scale of buildings also help in the orientation and identification of place. As we approach a building and make our way around and within it, we constantly need cues about direction and utility to help us on our way. The building itself can, and should, provide most of these cues. The ability to anticipate what is ahead prepares us psychologically and is the architectural equivalent to physical warming-up. The stress that stems from illegibility in design is not commonly acknowledged but surely adds to the aggravation of modern life.

4.4 Sleep, rest, & recuperation

All activities, desirable or otherwise, are stressful to the body. It is imperative that our energy is restored through sleep, rest and recuperation. The proper environmental cues - low lighting, low noise levels, amenable surroundings - are important for these restorative activities.

Anabolic processes, or body repair, occur only when the body is under low stress or at rest (McArdle, et al., 1991). The environment can help reduce the stress endured by the body. In particular, the presence of nature and loved ones can be restorative, unless of course, they are associated with the stress itself. For example, someone who has just lost his job may find any reminders of his family, and hence his responsibility to them, painful. As well, someone who has been demoted to a lower position in a smaller and rural branch may find nature scenes uncongenial. Low levels of noise and light are important in reducing the tension in the body and in enabling rest and sleep. At the same time, some sounds may be soothing. Some people may also find some colours soothing but other colours distracting and disturbing.

4.5 Stress & exercise

> " *The major objective in training is to cause biologic adaptations to improve performance in specific tasks. (Pg 423)*
>
> *A specific exercise **overload** must be applied to enhance physiologic improvement effectively and to bring about a training change. By exercising at a level above normal, a variety of training adaptations enable the body to function more efficiently. The appropriate overload for each person can be achieved by manipulating combinations of training **frequency, intensity, mode,** and **duration**. (Pg 424-425) "*
>
> McArdle, W D, et al (1991)
"Exercise Physiology"

There is some evidence to suggest that exceptional but controlled stress can be beneficial in inducing biological adaptations that improve the body's resistance to future stress. The imposition of adaptive stress levels must be carried out with care because of the extreme strain imposed upon the body. There must be adequate preparation and recuperation periods.

Physical exercise is useful in more than physical ways. It has been found that exercise increases not only physical strength and self-esteem but also promotes self-dependency and develops general coping resources (Long and Flood 1993). In other words, because stress provokes a general response, exercise prepares the body not just for physical stress but for other kinds of stresses as well.

Even light exercise can be beneficial if undertaken on a regular basis (McArdle, et al, 1991; Long and Flood, 1993). The benefits of stress, even at moderate levels, can be and is accumulative. Overcoming this level of stress prepares the body for other, perhaps more difficult, stresses. Regular exercise exposes the body to stress beyond the range normally encountered and enables the body to better withstand the occasional unexpected trauma.

Environmental stress can also be understood in this way. Exciting environments and exciting art provide more than moderate stimulation but, to be useful, such stimulation must be presented at a manageable level or within a manageable framework.   Art and leisure - including fine arts, architecture, film, television, computer games, and sports - provide such a framework for safe stimulating experiences.   We are aroused to levels that might otherwise be distressful and/or damaging if the framework within which these experiences occur do not exist to protect us from their ultimate reality. Because the framework does exist, we are able to enjoy the stimulation and also to learn more about ourselves, about the environment and about life, through these experiences.

5 Stress & Thermal Experience

As noted above, the direct effects of heat or cold are not in themselves stress symptoms.   The occasion of sweat or shivering is certainly a response to thermal changes but these reactions are not exacerbated by other stressors. It is unlikely that the subject will sweat more if exposed to strong winds, for example. What is more likely is that the irritability of the subject would be increased as a result of coping, or trying to cope, with additional stressors.

This distinction is important because except for extremes of temperatures, our ability to adapt to heat or cold is often enough to maintain a reasonable level of comfort (Humphreys 1975).   Imagine, for a moment, that you are working in an environment that is slightly too cold for comfort. What happens next? Rather than continue to work in discomfort, you would adapt to the situation. You might have a cup of coffee or tea, you may rub your hands together and work a little harder, you may even leave the room in search of warmth (by a fire or heater, perhaps).   You might close the window if it was open or put on a jumper if one was handy. If you could, you might turn the thermostat higher. You may even decide to finish off early. All these incidental adaptations are not of direct consequence to your health nor to your satisfaction with the space.   What is of consequence is if these avenues are somehow absent. If, for example, you are cold and cannot huddle in a corner nor put on a jumper but have to concentrate on the demands of the job at hand.

It is perhaps this distinction between the accumulative effects of stress and thermal adaptation that is at the root of our inability to develop an adequate standard for the thermal environment. If adequately allowed for, adaptation to changes in the thermal environment be it air temperature, solar radiation, wind or humidity, is not problematic. What is problematic is when these adaptations are not catered for and the mismatch between the user's needs and the provisions in the environment adds on to the other stressors she has to cope with.

The result is not a user furiously shivering while she works but one who is irritable and mistake prone. When questioned, she is more likely to blame her job, her boss or the general work environment than to identify the thermal culprit. Quite rightly, she would place her work and her human relationships of higher importance than the environment.   And quite rightly as well, she would measure the environmental provision, or lack of it, as indicative of the concerns of management and hold the management generally responsible.

Without the pressure of other stressors, we are more able to adapt our behaviour to the environment and will be generally more satisfied. Going to the beach, having a casual swim, taking a morning jog, reading a book recreationally are activities that are not themselves highly stressful and for such activities, we do not often complain about the environment. On the other hand, for professional swimmers and athletes or editors who have to meet datelines, the environment in which they compete or carry out their work is of much greater significance.

The upshot of this is that for the workplace, and for the kind of environments normally encountered in buildings, designing for thermal comfort in the sense defined by Fanger has no practical value. Human needs and the range of activities engaged in within any particular place are so wide as to require the provision of a user-controllable range of thermal conditions. The provision of such controls need not be technologically demanding. Instead, as the examples of windows, fireplaces, conservatories, and other vernacular archetypes show, good *places* often have this facility to provide different thermal conditions within one location or else allow for easy modification.

5.1 The singular environment

It is a mistake to assume that the division of the environment into heat, light and sound is applicable in practice. Our discussion of stress indicates that thermal stress is experienced not in classical thermal terms of sweating or shivering but in terms of irritability and frustration. These emotions are a result not of the thermal conditions themselves but of the inability to respond adequately to those conditions. We have argued that the thermal stresses experienced in buildings are of the kind that adds to an already stressful situation rather than constitute a uniquely definable stressor.

In the same way, stresses from noise and light - just to name the other two major concerns of environmental science - of the sort common in most buildings do not manifest themselves in a manner that would immediately indicate their source. The effects are symptomatic of general stress - irritability, proneness to accidents and mistakes, minor ailments like headaches and colds, and depressed moods. And the best cure may be simply not to continue with the stressful activity rather than to change the environment.

This last option is not one that is meaningful to our context, however. As environmentalists, as opposed to management or politicians, say, we must assume that there are good reasons for wanting to engage in these stressful activities and our responsibility must be to reduce additional stresses from the environment. Fortunately, there are some research which suggest that productivity can be improved environmentally - with the provision of background music, bright lighting, suitable temperatures and 'fresh' air (Penn & Bootzin 1990, von Restorff, et al, 1989). Notice that the criteria are subjective - what is music to some might be noise to others; *bright* lighting and *suitable* temperatures fall into that region of uncertainty as to be undefinable; and 'fresh' air is objectively non-sensical since the air we breathe has been with us for millions of years and is only recycled by ecological processes. 'Fresh' air is probably characterised by smell, low temperatures, air movement and by physiological reaction (whether or not the nose blocks up and headaches) rather than oxygen content. Indeed, von Restorff, et al, found no correlation between mental productivity and a reduction in air oxygen (15% as opposed to the normal 21%) and a corresponding increase in carbon dioxide (5% vs 0.03% normally). If, as von Restorff, et al, suggest, we can survive at less optimal O_2/CO_2 concentrations than currently normal, then it stands to reason that we are less sensitive to O_2/CO_2 concentrations than we are to pollutants and air particulates - the presence of which would be signalled, if at all, by smells and irritation to the nose and headaches. Notice also that there are more than one requirement and that the benefits of one (music, say) can be negated by the distress caused by another (high temperatures, say). Consequently, it must be added that the parameters listed are not unequivocal. Nevertheless, to the list I would add the need for a view of the outside - with an emphasis on vegetation and sky (Kaplan & Kaplan 1989, pg 1).

Thus, while it may be scientifically convenient and experimentally rigorous to adopt a reductionist and separatist approach to the study of environmental science, such an approach cannot be defended during design. Given the level and nature of environmental stress in buildings, environmental design needs to take a broader and more sophisticated view.

6 The Sensorium

There is a growing body of research on the social and cultural determination of sensory perception (Howes 1991, Scientific American 1976). Much of the work has been concerned with the visual and auditory senses. This is not surprising as they form the basis of human communication - writing, art, speech & music. In this field of research, the senses are often treated as a whole; a *sensorium*, and there is an emphasis on the interrelationships between the senses and across cultures and societies.

It appears that the division of our senses into sight, hearing, taste, smell & touch is culture based. Some cultures recognise as few as two different senses while others recognise phenomena, like speech, as sense - which we would not consider as one. Even more telling, the amoeba has no classifiable neural tissue let alone sense organs but is nevertheless sensitive to chemical changes in its environment as well as to temperature, mechanical, gravitational and electromagnetic stimuli (Christman, 1979). Human babies (Maurer & Maurer,1988) exhibit synaesthesia but most lose this ability upon maturity. A review of literature on the evolution of the senses does not produce a clear lineage of the development of individual organs but rather a history of development of sensory perception that is in the amoeba embedded in the whole organism, to specialised receptor organs in multicellular and radially symmetrical animals like the jellyfish, to the development of bilateral symmetry with a brain and subsidiary nervous system.

Our mind and our senses interact in subtle and little understood ways to give us a coherent understanding of the world around us (Marks, 1978). In some instances, our senses provide the same information. For example, we can both *see* that an object is sharp and *feel* that it is sharp against our skin. The *sound* that it makes as it scrapes across a surface also tells us of its sharpness. We can also tell how far away something is from seeing it and from hearing it. Lawrence Marks identifies size, form, space, time, motion, and number as qualities or properties that we might discover equally through more than one of our senses. Another interaction between the senses, according to Marks, is the similarity in the properties of the senses themselves. Thus, colours can be loud (like sound), sounds and colours can be soft (like touch), and smells can be sharp (like touch). The reason why we are able to use some properties of one sense interchangeably with another is because the senses share common qualitative measures. These common qualities are extension, intensity, and brightness. The interchangeability and correspondences between the various senses is not surprising if we realise that our senses and our responses to them is not specific responses to specific sensations but aspects of an overall behavioural pattern of survival and adaptation to the physical environment. In all, Marks identified five doctrines common to all the senses:

1. the doctrine of equivalent information
2. the doctrine of analogous sensory attributes and qualities
3. the doctrine of common psychophysical properties
4. the doctrine of neural correspondences
5. the unity of the senses

The point of this foray into the sensorium is to underline the fact that sensations like heat, light and sound are not merely background conditions for the performance of various activities. Rather, they provide cues and information about the environment and suggest corresponding behavioural patterns in the recipient. The gross simplification of the sensory environment in environmental science into content-free measures of background conditions has led to the exclusion of useful sense-data and the inadvertent substitution of meaning-less sensation for useful information.

6.1 Cultural influences on thermal perception

There has been several studies on cultural adaptations to the thermal environment (Rapoport 1969, and Humphreys 1975 & 1995, for example). From these studies, one might expect that different cultures would have different perceptions of the thermal environment. Conventional wisdom associates skin colour, height and shape of different peoples with adaptation to climate. Such beliefs are however disputed by Fanger (1970) who argued that no statistically significant difference was found in his studies that could be attributed to age, sex or race alone. The explanation that lies behind Fanger's findings is that our bodies are so physiologically similar that the differences within each group of age, sex or race is sufficient to counterbalance the differences between the groups.

In contrast with Fanger's view is a report on feral or wild children. Classen (1991) reviewed the case studies of three such children. Victor was a wild boy found in the woods of Aveyron, France, in 1800 AD. He was thought to be about 12 years old when found and might have been living in the woods for about 6 years. The boy was then taken to Paris where he was studied, cared for, and taught. Kamala was the elder of two sisters found in a wolf den in India in 1920 huddled together with 2 cubs and a mother wolf. The younger girl died a year after discovery while Kamala survived for nine more years. Kaspar was found at the age of 16 in Nuremberg in 1828. When taught to speak, Kaspar related a childhood spend in solitary confinement, unexposed to changes in weather and the environment, and fed only with bread and water by an unseen caretaker. All three children showed abnormal sensitivities. In many cases, their senses were more acute than normal - they could smell apparently odourless objects like pebbles, or distinguish fruit trees from a distance by smell; Kaspar could distinguish certain metals by touch alone and whether a magnet is pointing its north or south poles towards him. But it is the thermal behaviour of these children that interest us most (pp 53-54):

" *Victor and Kamala were indifferent to heat and cold. Victor was*

*able to pick up and eat boiling potatoes, and Kamala went naked in
the chill of winter with no ill effects. They had to be conditioned
into feeling differences in temperature. Kaspar, having lived in an
environment of uniform temperature, had no prior experience with
heat and cold and was extremely sensitive to both. The first time he
touched snow, in fact, he howled with pain. Victor, in contrast,
rolled himself half-dressed in the snow with delight. "*

We can only speculate to what extent our overly protective environments have
undermined our resilience to the weather and how much this lack of sensory
stimulation, or exercise, contributes to the increase in minor complaints like colds,
headaches and so on.

7 From Homogeneity to Heterogeneity

The above discussion attempted to show that the biophysical explanation of stress is
useful in developing a greater understanding of observed environmental behaviour.
Given the tools available today and the level of research and training now common, is
it still necessary to retain a single number criterion for thermal design? Perhaps a
more complex and sophisticated approach to environmental science is now possible.

A complete restructuring of environmental science cannot be addressed within the
scope of this paper. The following framework is presented purely as a basis for
discussion and further development.

7.1 The general theory of stress

We have tried to show in this paper that research in stress can provide us with a
general understanding of the mechanics behind the environmental concerns of heat,
light and sound. In particular, we would like to note the following:

1. The effect of environmental stress depends on several factors including the
   emotional and physical state of the individual.
2. The transition between different environmental experiences can be assisted or
   hindered by design and contribute directly to the degree of stress experienced.
3. Environmental sensations like hot/cold, bright/dark, noisy/quiet are not neutral
   measures of environmental stimulation but are cues to the level of activity and
   alertness required to function in the environment.
4. The human body operates best with periods of high activity followed by periods
   of rest and low activity.
5. There is a need for stimulating environments just as there is equally a need for
   neutral and restful environments.
6. The environment is sensed as a whole and the effect of the individual sensation
   (heat/light/sound) is affected by the context in which the sensation is
   experienced.
7. Responses to the environment are culturally conditioned.

7.2 Design as mitigating experience

Emotions evoked through art are less stressful than real events. We are usually able to put aside the feelings, however passionate, aroused through art (like movies, for example) much more easily than we can if the same events happen to us in real life. In terms of our stress model (Fig.3), what art does is to raise the pertinent thresholds. This happens both because we engage with art mentally rather than physically *and* because art provides a *greater*, yet safer, range of sensory and emotional experiences. Art thus works as psychological and sensory exercise.

The environments formed by good architecture often have this mitigating quality of art. The glory of nature is presented without exposing the user to its dangers. It is significant that to most of us, the word *shelter* is synonymous with *building*. This is true even though we can find shelter under tress, behind rocks and inside caves, and buildings can be a source of pleasure (the home), pride (civic buildings, palaces) and fear (prisons). For most of us, the provision of a bed in a room is not enough for rest and recuperation. We often need to *go home*, to gain as well a sense of belonging, quiet and tranquillity in order to feel restful.

Architecture is essentially about protection (or *shelter*), of course. But architecture can also provide stimulating environments. *Places* that we enjoy and often return to usually have this stimulating quality which we find restorative. If we stop to consider why we enjoy these places, it is often because of the *variety* of ways we can interact with the environment and the sensory *stimulation* we can find there. Good *places* not only *protect* us but also *enhance* our experience of the environment.

7.3 The emotional impact of environment

Lighting design today no longer advocates the homogenous lighting of the mid-sixties. More than just a source to facilitate visual tasks, light patterns(Flynn 1988, pg 157) affect personal orientation and user understanding of the room and its artefacts: *"Spot-lights or shelf-lighting affects user attention and consciousness; wall-lighting or corner-lighting affects user understanding of room size and shape. Considered as a system, these elements establish a sense of visual limits or enclosure."* Psychological impressions that can be achieved through lighting alone include sombreness, cheerfulness, playfulness, pleasantness, tension, intimacy, privacy and warmth.

The thermal environment is similarly full of psychological implications. The thermal engineer has more elements to play with - radiant heat, air temperature, wind and humidity. At the user end, she can feel warm, cosy, chilled, cool, cheerful (as in a *cheerful* fire or to be *cheered* by a fire), aglow, sunny, or fresh and keen - to name but a few. The ways by which the thermal environment may achieve such effects await further study.

7.4 Place typology

If the environment is not perceived in neutral terms but as specific settings for particular activities, it follows that there is a need to develop environmental research into typologically specific studies. In doing so, there is a need not to be too narrow in the definition of the typological function but to attempt to encompass not only the different configurations common to the typology but also the range of activities that might be found.

If we consider office design as a typology, for example, one of the first points we need to make is that good office design must be concerned with the ancillary spaces as much as it is concerned with the primary workplace. The work phenomenon must be regarded as a whole: the ritual from getting up to a preset time; getting to work; relating both professionally and socially in the course of the day; taking breaks - coffee, lunch; going home; expectations at home; and the five-day working week. Looked at from that viewpoint, the office workplace represents an important and primary element in the office-worker's life. As such, the workplace must be designed to provide good places for work, respite & stimulation. The office workplace is not just a place where we go to work and then go home but a place where we spend a major part of our adult life, where we contribute to society, where we most feel its pulse, and for which we have worked hard in school. The office workplace help define who we are, what we do and where we stand in society. We spend more of our waking hours at work than we do at home. Our needs for the office are therefore at least as complex as those that we have for our homes.

The purpose of making typological studies is not to provide a straightjacket to design possibilities but to ensure that the research applied is indeed relevant and useful to the design brief. The task is to reconcile existing research on any particular design type and to fill in the gaps that narrowly-defined individual research cannot address. Both physical and psychological factors must be considered. In this regard, it is important to develop a historical and anthropological understanding of the way the building type has been used and to attempt to evaluate how its future use might change.

The typological division of buildings is admittedly problematic. Buildings notoriously are often adapted for uses not allowed for in the original brief. One typological ordering that may have seemed appropriate at any one time may become suddenly inadequate subsequently. Nevertheless, environmental science research must be deemed to have reached a stage where such division of our discipline is much needed.

7.5 Parametric studies

The existing building stock provides a wealth of research potential but even so, there is a need to work out the limits to which existing technology and know-how can be stretched. There is a need, in other words, to carry out theoretical parametric studies.

Sophisticated modelling tools exist but they have yet to be used to *exhaustively* explore various building configurations. Such studies will help us appreciate the boundaries within which research or theory can be applied. Furthermore, as student, post-graduate and research projects, they will help professionals appreciate the complexity of the design environment.

7.6 Design aids

Sophisticated design tools which enable designers to see the implications of their design and thereby match it against a complex set of criteria rather than a single standard that oversimplifies research findings are now available. Having produced a particular design, we can now subject the design to different environmental conditions and see how it performs. As the environmental conditions change, we can also incorporate features into our design that adapt to these changes and not be restricted to designing fixed building elements.

Such design aids need to be in more common use and to be developed further to make them both more sophisticated and more easy-to-use. With such aids, we no longer need to design to oversimplified and inappropriate criteria but can tailor our designs to the whole range of complex human needs.

7.7 Heterogeneous criteria

The word *heterogeneous* stems from the Greek *heteros* meaning other. The adjective suggests a composition of many parts and of different kinds (Chambers 1983). There is much environmental research to suggest that our environmental needs are much more complex than we have been able to embody in our standards. Perhaps we should try to develop standards that reflect our research rather than an outdated criterion of simplicity that need no longer apply given today's computing power.

In opening up environmental criteria to allow complexity, we must be careful to maintain accessibility so that others may build upon, criticise and otherwise improve upon existing work.

8 Conclusion

This paper has used stress as a basis for explaining thermal behaviour. Stress is a result of the mismatch between our internal environment and the external and reflects our homeostatic response. The intensity of stress depends on many things, all of which change between people and with time. To reduce stress, we should not simply reduce environmental stimulation (although this is to some extent necessary) but also design for heterogeneity so that users may seek the environment that suit them most at any particular time.

This approach, using stress as explanation and aiming for heterogeneous design, provides a basis for the development of a more sophisticated environmental science. Given our present state of art, we ought to view our research in all its complexity and expand our field to include other research that can help us understand better the mechanics of environmental response. There is no longer a need to be constrained by a strictly reductionist approach. This is particularly so when observations cannot be reconciled with theory.

8.1 Acknowledgements

In closing, I would like to acknowledge the critical assistance of Dr David Crowther at the Martin Centre in the development of the stress explanation presented here. The comments and criticism of Dr Nick Baker of the Martin Centre, some of whose papers are referred to here, have been taken to heart.

Most particularly, I would like to acknowledge the inspiration and encouragement of my supervisor, Dr. Dean Hawkes, and for blazing the trail upon which I now tread. The ideas presented here are my responsibility but are derivative of his own views, his work and his beliefs. In particular, the notion of developing environmental studies based upon typological and parametric investigations arose from his comments during supervision.

9 References

1.   ISO 7730 *Moderate thermal environments - Determination of the PMV and PPD indices and specification of the conditions for thermal comfort* First edition 1984-08-15 UDC 331.043.6 Ref No ISO 7730-1984 (E)
2.   Baker, N V (1994) *The Irritable Occupant* The Martin Centre Research Society Lectures
3.   Baker, N V (1995) Adaptive Opportunity as a Comfort Parameter *Workplace Comfort Forum Proceedings* 22-23 March RIBA London
4.   Borysenko, Joan (1984) Stress, Coping, and the Immune System *Behavioral Health: A Handbook of Health Enhancement and Disease Prevention* Matarazzo J D, Weiss S M, Herd J A, Miller N E, & Weiss S M (eds) John Wiley & Sons New York pp 238-260
5.   Cannon, Walter B (1932) *The Wisdom of the Body* Norton New York
6.   Chambers 20th Century Dictionary 1983 edition
7.   Christman, Raymond John (1979) *Sensory Experience* 2nd ed Harper and Row New York London
8.   Classen, Constance (1991) The Sensory Orders of 'Wild Children' *The Varieties of Sensory Experience* Howes, David (ed) University of Toronto Press Toronto London pp 47-60
9.   Cooper, Ian (1982) Comfort Theory and Practice: Barriers to the Conservation of Energy by Building Occupants *Applied Energy* Vol 11 pp 243-288
10. Cousins, Norman (1979) *Anatomy of an Illness as Perceived by the Patient* W W Norton & Company New York and London
11. Evans, Gary W (1982) *Environmental Stress* Cambridge University Press
12. Fanger, P O (1970) *Thermal Comfort* Technical Press Copenhagen
13. Flynn, John E (1988) Lighting-design decisions as interventions in human visual space *Environmental Aesthetics* Nasar, Jack L (ed) Cambridge University Press Cambridge UK pp 156-170
14. Glass, R C (1977) *Behavior Patterns, Stress and Coronary Disease* Lawrence Erlbaum Hillsdale New Jersey
15. Hawkes, Dean U & Willey, H (1977) User response in the environmental control system *Transactions of the Martin Centre for Architectural and Urban Studies* Vol 2 pp 111-135 Woodhead-Faulkner Cambridge
16. Hawkes, Dean U & Haigh, Diane (May 1980) *User Response in Environmental Control* Martin Centre Technical Report
17. Hawkes, Dean U (1988) "Environment at the Threshold" in *A New Frontier: Environments for Innovation* (Proceedings: International Symposium on Advanced Comfort Systems for the Work Environment) by Knoner, Walter M (ed) May 1988 pp 109-117, Center for Architectural Research, Rensselaer Polytechnic Institute, Troy, New York - to be reprinted in Hawkes, D U (1995) *The Environmental Tradition: Studies in the Architecture of Environment* E & F N Spon London
18. Hawkes, D U (1992) The Language Barrier *Architect's Journal* 5 February pp 22-25
19. Hawkes, D U (1995) *The Centre and the Periphery* The Martin Centre Research Society Lectures - to be published in *Architectural Research* Vol 1 No 1

20. Humphreys, M A (1975) *Field Studies of Thermal Comfort Compared and Applied* Building Research Establishment Current Paper CP 76/75 August

21. Humphreys, Rev. M A (1995) *What causes thermal discomfort* Workplace Comfort Forum 22-23 March RIBA London

22. Jokl, Ernst (1977) The Immunological Status of Athletes *Medicine Sport* Vol 10 pp 129-134 Karger Basel

23. Kerr, J H; and Vos M C H (1993) Employee fitness programmes, absenteeism and general well-being *Work & Stress* Vol 7 No 2 pp 179-190

24. Long, B C and Flood, K R (1993) Coping with stress: psychological benefits of exercise *Work & Stress* Vol 7 No 2 pp 109-119

25. Marks, Lawrence E (1978) *The Unity of the Senses: Interrelations among the Modalities* Academic Press New York & London

26. Maurer, Daphne & Maurer, Charles (1989) *The World of the Newborn* Viking London

27. McArdle, W D; Katch, F I & Katch, V L (1991) *Exercise Physiology: Energy, Nutrition, and Human Performance* Lea & Febiger, Philadelphia/London

28. Newsham G R & Tiller D K (1995) Thermal Comfort in Offices: A Study Using a Computer-based Questionnaire *Proceedings 26th Annual Meeting of the Environmental Design Research Association* EDRA Boston

29. Ong, B L (1994) Designing for the Individual: A Radical Interpretation of ISO 7730 *Standards for Thermal Comfort* Conference, Windsor - published in *Standards for Thermal Comfort: Indoor air temperature standards for the 21st century* Fergus Nicol, Michael Humphreys, Oliver Sykes & Susan Roaf (eds) E & FN Spon London 1995

30. Ong, B L (1995) *The Biology of Beauty* The Martin Centre Research Society Lectures - unpublished

31. Penn, Patricia E & Bootzin, Richard R (1990) Behavioural techniques for enhancing alertness and performance in shift work *Work & Stress* Vol 4 No 3 pp 213-226

32. Rapoport, Amos (1969) *House Form and Culture* Englewood Cliffs/Prentice Hall New York

33. Scientific American, readings from (1976) *Recent Progress in Perception* with introductions by Richard Held and Whitman Richards W H Freeman & Company San Francisco

34. Selye, Hans (1973) The Evolution of the Stress Concept *American Scientist* Vol 61 Nov-Dec pp 692-699

35. Selye, Hans (1976) *The Stress of Life* 2nd ed. McGraw-Hill New York London - original edition in 1956

36. Thayer, Robert E (1989) *The Biopsychology of Mood and Arousal* Oxford University Press Oxford

37. von Restorff, W; Kleinhanss, G; Schaad, G and Gorges, W (1989) Combined work stresses: effect of reduced air renewal on psychological performance during 72h sustained operations *Work & Stress* Vol 3 No 1 pp 15-20

CHAPTER THREE

Specifying Indoor Climate Data

Derek John Clements-Croome

Department of Construction Management and Engineering,
The University of Reading, Reading, UK

1 Summary

Buildings are designed to suit the climate in which they are located and the functions for which they are intended. There is a unique relationship between an individual, the environment and the building they inhabit. Everyday experiences tell us that there are a host of factors which are relevant to this concept. Not only do air and surface temperatures, humidity, air movement and air purity play a part, but psycho-sociological factors also have an important role. The attitudes of people around us, the organisation of space, colour schemes and many other factors all can have an influence on our mood and work output. Since there is an interaction between all these factors, the problem is complicated further. A deficiency in one of the physical factors can spoil the balance of the environment. Equally so, surroundings which contain disturbing social or psychological aspects can be uncomfortable.

Environmental stimuli are sensed and transmitted by the nervous system to the brain. Depending on your viewpoint the brain and the mind, or the brain-mind as an entity, respond accordingly. This response has as inherent characteristic originating in the genetic code and the subsequent environmental conditioning of the individual, and a transient dynamic characteristic which depends on the adaption capacity of the individual to a changing environmental scene. A sense of well-being for a particular person requires healthy mind and body. Mind and body are related to one another via the hormone system, the pattern of which determines mood and ultimately a sense of well-being. Well-being is important for good work production. Task performance is best when the mind is alert at an optimum arousal level with the lease amount of distraction assuming the hormone and immune systems of the body are working effectively.

Naturally Ventilated Buildings: Buildings for the senses, the economy and society. Edited by D. Clements-Croome. Published in 1997 by E & FN Spon. ISBN 0 419 21520 4

Comfort is much more confined in context than well-being. A comfortable environment is one in which there is freedom from annoyance and distraction, so that working or pleasure tasks can be carried out unhindered physically or mentally. A building and its environment can help people to produce better work because they are happier when their minds are concentrated on the job in hand and building design can help this to be achieved. Boredom and lassitude, or anxiety and drug-induced mental states represent two ends of a spectrum which counter the optimum balance point of this attention or arousal level continuum. At low and high arousal levels the capacity for performing the work task is low: at the optimum level the individual can concentrate on the work task while being aware of the peripheral stimuli from the physical environment. Different work tasks need different environmental settings.

We need to assess if a comfortable indoor environment is a necessity for the occupants good health and high productivity. We may need to redefine comfort in terms of well-being. There are three current standards providing guidance for the assessment of occupant comfort: ASHRAE standard 55-92 [5]; ASHRAE Standard 62-89 [4]; and ISO Standard 7730 [58]. They all emphasise thermal comfort rather than overall comfort. Based on Fanger's PMV/PPD model, the international thermal comfort standard ISO 7730 has been updated, but has not yet been approved as the new pr ENV 1752 *Ventilation for Buildings: Design Criteria for the Indoor Environment*. The PMV/PPD model has also been used in computerised fluid dynamics to predict the difference in thermal sensation in the occupied zone of spaces. However, field studies show temperature effects which are not found in climate chambers. Measured laboratory and field neutral temperatures vary by as much as 4°C or more. Field researchers argue that other variables maybe the cause of such differences. The PMV/PPD method does not consider any effects due to adaptation, cultural differences, climate and seasons, age, sex differences or psychological attributes due to expectations and attitudes; in addition effects of thermal conduction through seating and effects of posture are ignored. These differences are even more important for naturally ventilated building which only offer a coarse level of temperature control compared with airconditioned ones.

The effect of an environment at any moment is dependant on ones past experiences, the clients at the time, and so a time sequence analysis is important. People are not passive recipients of their environment, but adapt physiologically and behaviourially. People do, however, adapt according to expectations and preferences. Some people are very sensitive to their environment; others are not so easily influenced by their surroundings. Clearly, each person has their own range of comfort values in different activity situations. People expect their surroundings to allow them to pursue their work activity unhindered. The task of the environmental engineer and the architect is to create environments which are acceptable to most of the people living inside buildings, but to allow us much individual control as possible.

2 Background

Standards aim to set acceptable conditions for various combinations of circumstances. A broader level of understanding is needed when interpreting the word *acceptable* so as to reflect not only the scientific basis of environmental specification but also cultural and social aspects [117].

The language of sustainable buildings is about climatically sensitive structures; it is about flexibility and adaptability. Often design standards are too fixed. Criteria could be more malleable but as Shove [117] points out this requires an approach which can focus upon cultural variability of building occupants and upon their creative, multi-dimensional interaction with the built environment. Patterns of energy consumption for example depend on people's decisions which are rooted in attitudes and actions. Ultimately a series of decisions, some of them become habits, constitute a lifestyle. This brings to mind the objective knowledge model of Karl Popper which relies on an interaction between social and economic as well as technical issues. People, pounds and principles are all important.

The difficulties of specifying the indoor climate of buildings arise from many causes. The obvious one is that there are many factors contributing towards the indoor environment and therefore the human response is difficult to assess in establishing design information for one particular variable, such as temperature or air movement. It is perhaps worth considering the reasoning that underline the various scientific methods [25].

Postulation involves producing particular solutions from general principles. For example, reducing temperature levels for buildings around the world from a mass of field data would involve this deduction method. In contrast *empiricism* involves analysing experimental results and by induction making some general conclusions. For example, the basis of ISO 7730 is the work of Fanger carried out in the laboratory at Lyngby in Denmark [46]. *Hypothetical modelling* involves taking artefact properties and simulating aspect of natural phenomena. *Taxonomy* is the system of classifying by systematic comparison and differentiation. *Statistical analysis* allows patterns and trends to be identified by contingency and uncertainty with respect to individuals of a population. All of the field and laboratory work referred to in this Chapter will have had statistical analysis applied to the results that are being investigated. Lastly, *genetic or historical derivation* uses history and developments in a particular field to explain similarity and diversity. This method has application in our field, because there are many examples over the centuries which explain how people have adapted buildings to cope with the environment in different regions of the world. All of these reasoning methods have some application in the field of environmental design, but empiricism and postulation supported by statistical analysis have predominated. It is essential to understand the reasoning methods when comparing the data from different sources.

Sir Dai Rees [104] asked the question: what is it in our genes and biochemistry and their interactions with environment which sets human beings apart from the rest of creation? How can psychological factors influence health? This poses the challenge of unifying psychology and physiology. Standards will be limited until this problem is understood.

In contrast to the sculptor or composer who mainly work alone, the architect designs buildings in collaboration with the clients and a large team of people. To be successful there needs to be shared aspirations, values and vision otherwise any creativity or motivation for the project will be stunted. For example, if only the engineer gives priority to green issues it might be difficult to persuade the other members of the design team to share that vision. Hopefully, by coherent dialogue and team learning, the group effect will be greater than that of the individuals and the impact of this on the resulting building will become clear. Individuals need to have core techniques to deal with problems, but if buildings are to be more than something utilitarian there needs to be some intuition and interpretation of needs and knowledge. Ultimately, *systems thinking* is paramount. The well-being of an individual in the workplace depends on personal circumstances, the organisation, as well as the buildings and the associated systems. It is important to understand the behaviour of the system even if particular elements in the system cannot be fully explained. The basic system we have to deal with is one that links climate, building exterior, building interior and systems, and their interaction with people. Clearly, there has to be feedback or control between the people and the various parts of this system.

The principal advantage of empirical methods are that laboratory tests can be carried out in a controlled environment. This usually means that a number of variables are kept constant while one is allowed to vary. The problem here is that this is unrealistic of what happens in a real life situation. People are exposed to a whole array of environmental stimuli and individual control of any of them is quite difficult, but the human system adapts over a fairly wide range of variation. It might be that it is the holistic pattern of environmental parameters which is more important than particular ones, except under particular circumstances. In a computer suite the temperature level is much more critical than in a general office space. In a television studio the background noise level is critical compared to more general spaces. Besides, short-term adaptation there is the effect of long-term acclimatization. An internal temperature of 28°C may be acceptable in warm to hot climates, but may be felt to be much too hot in more equitable ones.

In seeking individual responses to environmental factors it is usually assumed that the subjective scales are linear, whereas it is more likely that they are non-linear. At 16°C for example most people in a sedentary occupation in an office would find it far too cold, and likewise at 35°C most people would find it too hot. Judgements converge at extreme conditions, but in between these extremes the body is tolerant to a wide range of change. In critical situations this neutral range is smaller.

Culture involves a mixture of aspects covering acclimatization, expectancies and environment of a particular region. In the USA for example, office workers will often tolerate higher sound levels than in Europe. Aspects of dress, eating and drinking also affect how the human system responds to the environment.

3 Present standards

The basis of the proposed European standard on indoor climate (pr ENV 1752 *Ventilation for Buildings; Design Criteria for the Indoor Environment*) is based on the work of Fanger originally set out in his book entitled *Thermal Comfort Analysis and Applications in Environmental Engineering* published by McGraw Hill in 1970. There has been a steady stream of notable laboratory work emanating from his laboratory since that time and the key references to this are in the *ASHRAE Handbook on Fundamentals 1993 Chapter 8*. This proposed standard is not yet approved because it is felt that there is too much laboratory data that has not been fully validated in the field. However, that does not undermine the great advancement of knowledge we have by the work that has been carried out during the last few decades. The steady state energy model representing the heat exchange between the body and its surroundings usually assumes that the body is in a state of thermal equilibrium with negligible heat storage. Various thermal exchanges by sensible heat loss from the skin; evaporative heat loss and respiratory losses result in a total skin heat loss and from there some assumptions have to be made with respect to the clothing of the body covering most of the skins surface. Further assumptions have to be made with regard to metabolic heat generation for various activities. With all of these assumptions it is not surprising that there is some considerable variation between individuals with respect to judgement of temperature since the basic physiological responses differ widely for individuals. Whilst accepting that no single environment is judged satisfactorily by everybody, even if they are wearing identical clothing and performing the same activity, the comfort zone specified in ASHRAE Standard 55-92 is based on 90% acceptance or 10% dissatisfied. Fanger [47] related the predicted percentage satisfied (PPD) to the predicted mean vote (PMV) as follows:-

$$PPD = 100 - 95 \exp \ [-(0.03353 \ PMV^4 + 0.2179 \ PMV^2)]$$

where dissatisfied is defined as anybody not voting either -1, 1 or 0. A PPD of 10% to corresponds to the PMV range of + 0.5, - 0.5 and it should be noted that even with a PMV=0 about 5% of the people are dissatisfied.

Peoples sensitivities vary considerably. Age, adaptation, sex, seasonal and circadian rhythms, local thermal discomfort, radiant asymmetry, temperature gradients, air quality, thermal conduction, posture, all will vary from one individual to another.

The draft European standard (pr ENV 1752) prescribes a PPD for the human body as a whole and distinguished that from local discomfort. Three quality categories are described. Category A corresponds to less than 6% PPD; category quality B for PPD less than 10% and quality category C for less than 15% as judged for the whole body. Local discomfort criteria are then defined for draft, temperature gradient, warm or cold floors and radiant asymmetry. The difficulties become apparent because it is very problematic to try and define so many variables in such a precise manner especially in a naturally ventilated building. In a laboratory situation this is possible, but even then difficulties arise because of variations between subjects being tested. Nevertheless an attempt to try and derive a suitable standard for groups of people working in similar spaces was a laudable idea.

The following figures show the kinds of variation one can expect between Fanger's prediction and field tests. The first figure shows some results taken in various lecture rooms at the University of Reading [28].

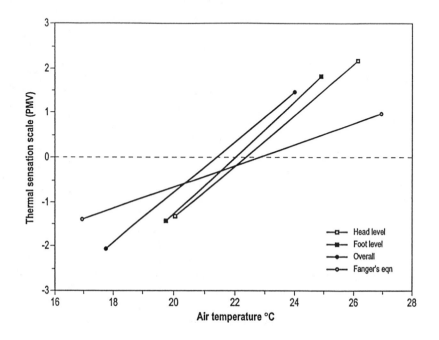

Figure 1: Effect of air temperature on thermal sensation responses

There is a notable difference between judgements made at head and foot level, and the slopes of the line between thermal sensation and air temperature are different for the field tests and the laboratory prediction. This becomes more evident above 22°C. In a more general context one can see by comparing the work of Gagge [48] and Fanger [46] with measured data by Brager [11] shown in Fig 2, that again there are significant differences between the field and laboratory data.

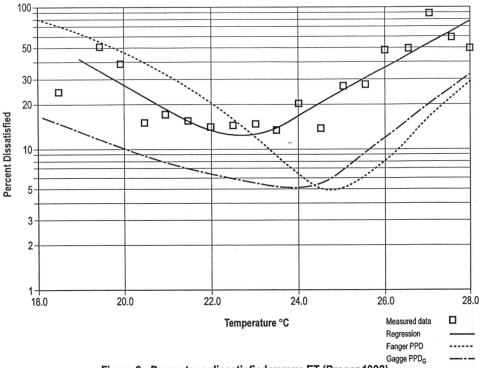

Figure 2: Percentage dissatisfied versus ET (Brager,1992)

The practical effects of these differences are shown in Table 1, where T_{il} represents the laboratory indoor temperature and T_{if} represents the field indoor temperature. T_o is the average annual outdoor temperature; conditions of 6°C and 10°C have been assumed. One can see that even at 1°C difference between laboratory predictions and field measurements this amounts to an energy saving of some 6-8%, whereas at a 2°C difference the amount of energy saving rises to between 12 and 16%.

$(T_{il} - T_{if})$ (K)	Energy Saving $(q_1 - q_f)/q_1$ (%)	
	for $T_o = 6°C$	for $T_o = 10°C$
0.5	3.0	3.9
0.8	4.8	6.3
1.0	6.0	7.9
1.4	8.4	11.0
1.7	10.2	13.4
2.0	12.0	15.7
2.4	14.4	18.9

Table 1: Energy Savings due to Lower Room Temperatures

The ratio of energy saving is given by:

$$(q_1 - q_f)/q_1 = (T_{il} - T_{if})/(T_{il} - T_o)$$

where T_{il} and T_{if} are the required room temperatures predicted from the laboratory data of Fanger and from the field measurements respectively.

Table 2 shows the influence of different clo values on neutral temperatures and hence the energy savings that can be effected. Clothing variations of 0.1 clo can make a difference of 4-5% in energy saving.

Clothing (clo)	Neutral temperature (°C)	Energy Saving (%)	
		$T_o = 6°C$	$T_o = 10°C$
0.8	22.5	0.0	0.0
0.9	21.8	4.2	5.6
1.0	21.1	8.5	11.2
1.1	20.4	12.7	16.8
1.2	19.8	16.4	21.6
1.3	19.1	20.6	27.2
1.4	18.4	24.8	32.8
1.5	17.7	29.0	38.4

Table 2: Neutral Temperatures and Energy Savings for Different Clothing Levels.

The assumptions made are air velocity = 0.076 m/s; metabolic rate = 1.2 met; vapour pressure of air = 1500 Pa; mean radiant temperature = air temperature.

Healthy buildings require adequate quantities of fresh air, but the precise amount of fresh air is difficult to estimate. A review of the historic work in this area is given by Croome [26]. Since then Fanger [44,45] has quantified air pollution sources by comparing them with a sedentary person in thermal comfort. The *olf* is defined as the emission rate of air pollutants from a standard person. Fanger generated curves that relate to the percentage of people dissatisfied with the emissions of one person in a laboratory chamber as a function of fresh air ventilation rate and obtained the following expression:

$D = 395 \exp(-1.83q^{0.26})$ for q > 0.332
$D = 100$ for q < 0.322

where:
   D = percentage dissatisfied persons
   q = ventilation-emission ratio, l/s x olf

A *decipol* scale has been derived where the decipol is the perceived air pollution in a space with a pollution source of one olf ventilated by 10 l/s of unpolluted air. Steady state conditions and complete mixing are assumed. On this basis healthy buildings are defined as those which have a decipol level of one and below, whereas sick buildings are defined as those which have a decipol value of about 8 to 10 or above.

A large number of people act as judges to sample the air in the environment. It may be that we have to return to some form of chemical analysis using chromatography as a basis for assessing the amount of fresh air required in a building in a more definite way. Nevertheless, attention has been drawn to the fact that not only people, but building materials and ventilation systems themselves contribute towards the pollutants in an environment. The effect of smoking has been well emphasised over the years, but importance in assessing emissions from building materials is a newer aspect that needs to be taken into account. With regard to the airflow system then maintenance is vitally important. In drafting pr ENV 1752 across many European countries, it became clear that the fresh air requirements in different countries varied considerably and also there still remains much work to be done to ascertain the correct amount of fresh air needed in spaces. Clearly, this is important for naturally ventilated, mechanically ventilated or airconditioned buildings.

4 Man & environment

Buildings are designed to suit the climate in which they are located and the functions for which they are intended. There is a unique relationship between an individual, the environment and the building they inhabit. Markus [80] describes the objective of a building in terms of client needs (see Fig 3).

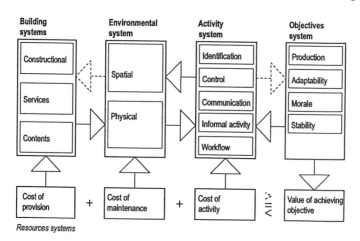

Figure 3: Conceptual model of the system of building and people (Markus *et al.*, 1972)

Morale and stability of the work force are important for good productivity which is a primary objective for the client which can only be achieved if the activity system of the organisation; the building with its constructions services and contents; the spatial and physical environment are all planned, designed, constructed and managed in an interdependent way so that the whole system works effectively.

It would be a better idea from the clients point of view if standards and data for building briefs reflected the effect of indoor environment on productivity. This Chapter will to develop this idea.

Environmental stimuli are sensed and transmitted by the nervous system to the brain. Depending on your viewpoint the brain and the mind, or the brain-mind, respond accordingly. This response has as inherent characteristic originating in the genetic code and the subsequent environmental conditioning of the individual, and a transient dynamic characteristic which depends on the adaptation capacity of the individual to a changing environmental scene. A sense of well-being for a particular person requires healthy mind and body. Mind and body are related to one another via the hormone system, the pattern of which determines mood and ultimately a sense of well-being. The immune system is another important aspect which is disturbed in sick buildings for example. Well-being is important for good work production. Task performance is best when the mind is alert at an optimum arousal level with the least amount of distraction assuming the hormone and immune systems are working effectively. This idea is summarized in Fig 4.

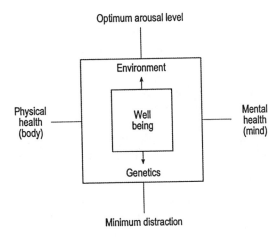

Figure 4: Defining Well-Being

It can be seen that bodily and mental health is an intrinsic pact of well-being but this is further enhanced by the human system being at an optimum attention or alertness level. Disturbance to the human system can arise from unfavourable interactions between people, environment and matter (see Fig 5).

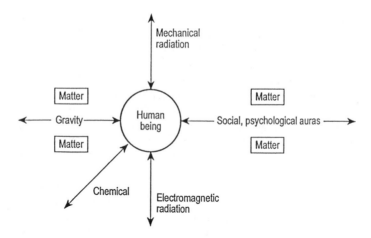

Figure 5: People, environment and matter

The human system is immersed within, and interacts with, the mechanical, electromagnetic, chemical and gravitational fields.

Buildings are a sensory experience. We dimension the world around primarily by light and sound. Most of the information entering the human system is via the eyes and the ears, but the senses of touch, taste and smell are also important. Concentration is the intense application of the mind resulting in complete attention; selective attention is the process by which information is selectively picked out. Good concentration leads to high productivity. To be alert is to be vigilantly attentive. The effort of attaining and sustaining concentration can be hindered or enhanced by the environment. A visual scene may be restful and aid concentration but a cluttered chaotic scene can be distracting.

Across the senses there is a physiological - psychological spectrum. Table 3 shows that stimuli induce organic reactions and can affect the physical and chemical barrier which form the entry to the innate immune system. The sensory organs process the signals and effect transmission to the brain. The long term memory selectively files sensory experiences and from these, associations and hence expectancies result. Choosing a restaurant for its ambiance, selecting interior decor or arranging an office are all actions which reflect individual sensitivities to environmental factors which can combine in various ways to make one feel good.

Partial Climate	Induces Organic Reactions	Stimulates Sensory Organs	Creates Associations	Describes the Environment
Light	*	*	*	*
Sound	*	*	*	*
Smell	*	*	*	
Heat	*	*		
Air Quality	*			
Elec. Climate	*			

Table 3: Climatic effects on the body and mind

Comfort is much more confined in context than well-being. Environmental comfort can be interpreted not only as an answer to physiological needs, but also as satisfaction with psychological and sociological requirements. A comfortable environment is one in which there is freedom from annoyance and distraction, so that working or pleasure tasks can be carried out unhindered physically or mentally. This interpretation of environmental comfort suggests considering all the partial climates which are linked to specific needs and to involve all the direct and indirect demands that influence well-being. Comfort must be interpreted as the real *science of living* denoting a state of complete satisfaction with the physical, mental, economic-social human conditions thus allowing people to fulfil their various activities in the building [108]. It is important to consider not only the building, but also all the systems and controls which are necessary for the operation of the building.

A building and its environment can help people to produce better work because they are happier when their minds are concentrated on the job in hand; building design can help this to be achieved. Boredom and lassitude, or anxiety and drug-induced mental states represent two ends of a spectrum which counter the optimum balance point of the attention or arousal level continuum.

DJ Clements-Croome

Fig. 6 is a three-dimensional model showing the effect of arousal state on the organization of mind capacity. The physiological and psychological processes through which ambient conditions influence individual reactions are arousal, stress, overload, distraction, and fatigue. At low and high arousal levels the capacity for performing the work task is low; at the optimum level the individual can concentrate on the work task while being aware of the peripheral stimuli from the physical environment. Different work tasks need different environmental settings.

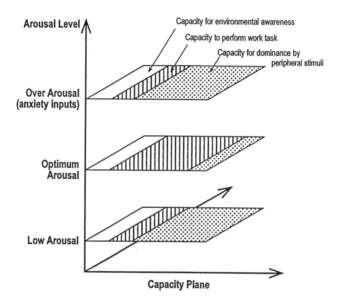

Figure 6: Relationship between arousal level and mind capacity

Fig. 7(a) shows that an optimum arousal level has to be attained for the best performance of work; this concept is known as the Yerkes-Dodson Law. According to the Yerkes-Dodson (1908), low levels of alertness will be insufficient to impact on performance whereas high levels of alertness will result in over excitation resulting in performance decline. Moderate alertness, it is suggested, will increase activation to an optimum level thereby improving performance. This relationship is moderated by task difficulty, with more complex tasks being performed better at lower overall levels of activation. Of course, individual differences in arousal level, information processing, and cognitive interpretations of the environment also moderate the relationship between arousal (stress) and performance [21,116].

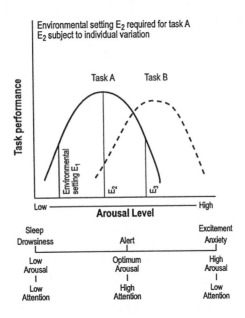

Figure 7: Different environmental settings are preferred for particular tasks

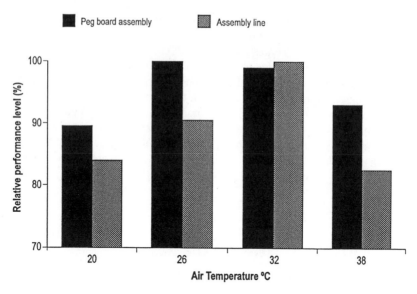

Figure 7b: **Relative performance in USA of two factory tasks at different temperatures (McNall, 1979)**

Fig. 7(b) presents a example of relative performance at different temperatures of two simple tasks. The assembly line workers show an optimum performance at 32 °C whereas the peg board assembly shows a maximum at 26 °C. Accordingly, when elements of the environment impinge on these capacities, it is expected that performance will decrease. Finally, fatigue is partly attributable to a physiological reaction to the environment occurring when muscles are overexerted, thereby resulting in decreased performance. The essential nature of mental fatigue is less understood.

Immersed is a physical environment, the sensory system in man has the opportunity to select many of the stimuli presented to it. Sensory pathways take the form shown in Fig. 8.

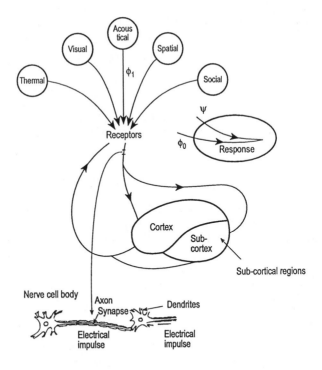

Figure 8: Sensory Pathways

The environment provides a stimulus input, φ_i, which is received by a receptor (e.g. ear, eye, thermoreceptor in the skin), where a transducer action takes place and the incoming signal is transformed from a sound wave, light quanta and temperature, or whatever, into a series of neural impulses which then travel through the central nervous system. Processing of the incoming data takes place in the cortex and sub-cortical regions of the brain after which a response occurs. The response to the environment may be an active or a passive one. The description so far has been a neurophysiological one, but the response has physiological and psychological components. James has suggested:

$$\varphi_o = f(\varphi_i, \Psi)$$

Where Ψ is a psyche function. Feedback, represented by the multi-variable φ_f in Fig. 8, occurs between the central nervous system and the sense receptors and is brought about by adaptation (a decrease in amplitude of the neurone discharge level for a given input) or fatigue. The sensory system may also undergo long-term adaptation in alien environments, called acclimatization.

Thus, φ_o can be expressed as :

$$\varphi_o = f(\varphi_i, \varphi_f, \Psi)$$

This function is a complex one and current research is aimed at obtaining a more complete knowledge about it. Classical psychophysics attempted to relate φ_o and φ_i in a simple way, the just-detectable change in any environmental response being related to a given fraction, k, of the input stimulus change:

$$\delta\varphi_o = K * (\delta\varphi_i/\varphi_i)$$

This relationship is only linear over a limited range of frequencies at certain loudness levels in the case of sound (Riescz 1928) and this is probable also true for thermal stimuli. Marshall [81] has proposed the concept of stochastic dominance in the central nervous system. An input signal is pictured as stepping from neuron to neuron with a constant probability of transition, P, along a sensory pathway. A probability function:

$$P_k = P (1-P)^{k-1}$$

is defined as the chance that an impulse will pass from one neuron to another in an interval of time k.Δt, assuming a transitional probability, P, common to all neuron inter-connections. Likely magnitudes of P and k for low and high levels of P_k are given in Table 4. This model permits all stimuli to enter the system and thus allows for the environmental awareness of the individual.

Some people are very sensitive to their environment - i.e. sensory inputs modify the sub-cortex field easily; others are not so easily influenced by their surroundings. Clearly, each person has own range of comfort values in different activity situations. People expect their surroundings to allow them to pursue their work activity unhindered. The task of the environmental engineer and the architect is not only to create environments which are acceptable to most of the people living inside buildings but also to try and cater for some individual variations.

Characteristic	P_k Level	Likely value of P and k
Load on Neural Network: e.g., If System Heavily Burdened	Low $(P_k \to 0)$	$(1-P)^{k-1} \to 0$ for $k \sim 1$, $k \sim 2$ P=0.1, P=0.9
Arousal State of Individual: (a) Drowsy (i.e. too low)	Low	k High (1 - 2)
(b) Optimal	High $(P_k \to 1)$	$\dfrac{\ln P \to k}{1 - \ln(1-P)}$ for $k > 0$, P>0.5 P=0.6 P=0.9 k=0.55 k=1.0
(c) Anxious (i.e. too high)	Low	k=high (1-2)

Table 4. Magnitudes of P and k for low and high levels of P_k [26]

5 Thermal comfort and adaptability

An extensive account of physiological principals and thermal comfort is given in the ASHRAE Handbook on Fundamentals 1993 Chapter 8.

Thermal comfort as defined by ASHRAE Standard 55 [5], is the condition of mind that expresses satisfaction with the thermal environment; it requires subjective as well as objective evaluation. ASHRAE Standard 55 and the International ISO Standard 7730 [58] are based exclusively on laboratory-based research using a steady state heat-balance model measuring the thermal interaction between and the surroundings the human body. The main comfort model is Fanger's Predicted Mean Vote (PMV) / Percentage People Dissatisfied (PPD) model [46]. The model allow prediction of comfortable temperatures in controlled environments but for naturally ventilated buildings (i.e. passive or free-running buildings) the model seems to break down. The PMV model has been interpreted as a constant set-point for given clothing and metabolic rate conditions for a specified air velocity. Recent research casts doubt upon the application of steady-state heat exchange equations to what is in practice a variable environment.

People are not passive recipients of the environment but take adaptive measures to secure thermal comfort. They can modify their clothing or activity; they can modify the environment such as the internal heat gains (lighting or solar heat gain) or modify the ventilation rate through opening doors and windows; they can also modify their posture [93]. In the longer term, people will seek a better climate or move to a more suitable one. Short-term and longer term feedback processes profoundly affect the relationship between a person and the thermal environment.

There is a conflict between the comfortable temperatures calculated from heat-exchange theory tested in the laboratory and those found in the conditions of daily life. The predicted mean vote (PMV) from Fanger's equation, have been compared with the actual mean comfort vote of the group of people and the discrepancies often found are equivalent to a difference on average of some 3 °C--4 °C too high. There are difficulties in applying the results of laboratory based thermal comfort research to the design of buildings [54].

The body expends energy the magnitude of which depends on the degree of work activity. This energy is derived from the oxidation of foodstuffs. Some will be used to satisfy immediate metabolic needs, some will be stored in the body, and some will be liberated as heat to the surroundings in order that the blood can remain at an almost constant temperature. Heat transfer between the body and its surroundings takes place by radiation, convection, conduction and evaporation.

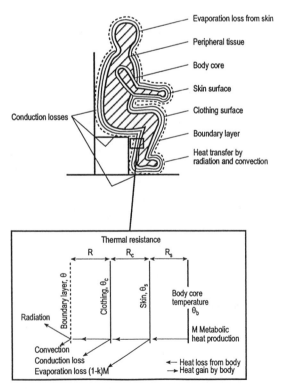

Figure 9:   The steady state heat exchange between the
            human body and its surroundings

These processes depend on the body surface temperature, the water vapour pressure between the skin and the air, the air velocity over the body, posture, clothing and surface area of the subject (see fig. 9). Croome [26] has described a simple way of considering the steady-state heat exchange between the human body and its surroundings which includes the effects of different metabolic rates and the weight of clothing. The temperature which is acceptable physiologically can be estimated using the equation:

$$\theta = \theta_b - M * [R_s + k * (R_c + R)]$$

Where θ is the comfort globe temperature; θ_b is the skin temperature; M is the metabolic rate; R_s is the skin thermal resistance; R_c is the convection conductance resistance; and R is the surface thermal resistance;

Using:

$$\theta_b = 37 \; ^\circ C; \; k = 0.7; \; R_s = 0.04 \; -- \; 0.09 \; m^2 \, ^\circ C \; W^{-1}$$

then:

$$\theta = 37 - M * [(0.04---0.09) + 0.7 * (R_c + R)]$$

There is considerable variation in the heat transfer coefficients between one individual and another. In hot arid climates the band of comfortable internal temperature θ is wider than that experienced in Europe [27]. The light clothing worn in the Middle East means that higher temperatures are acceptable physiologically and these can be estimated using the equation:

$$\theta = 37 - M * [0.05 + 0.7 * (R_c + 0.113)]$$

Thus:

$$\theta = 27.3 \; ^\circ C; \; assuming \; M=60 \; W/m^2, \; R_c=0.047 \; m^2 \, ^\circ C/W$$

Matching clothing, activity and temperature is the basis of the so-called adaptive comfort model. There is a substantial body of field data supporting this approach.

Humphreys [56,55] compared the results of field-studies of thermal comfort in many countries showed that different groups of people were comfortable at remarkably different temperatures [35] (See Fig. 10).

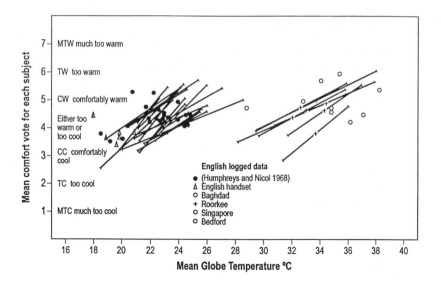

Figure 10: Comfort votes correlated with the mean
globe temperature (Dick, 1971)

The range of these differences was about twice that which could be attributed
to differences of clothing between populations. For example, one office survey
in Perth, Australia, among secretaries showed that the thermal conditions were
acceptable despite the fact that the mean indoor temperature was over 28 °C
[17]. Thermal comfort and acoustic surveys carried out by Wilson, Dubiel and
Nicol [133] shows that a comfort temperature of 30.1°C is acceptable in
offices in Pakistan, but a level of 23.3°C is acceptable in the UK. Similarly,
as regards noise there is a higher tolerance level by some 7dB in Pakistan,
compared to the UK. For a European these conditions would be unacceptable.
Thermal comfort surveys in homes and offices need to consider social and
behaviourial influences of the occupants and their perception of the thermal
environment as well as recording physical variables. Subjectivity varies and the
meaning of the words warm, hot, cool and cold vary around the world. When
comparing surveys from different climates or socio-economic regions it is
important to consider the influence of local conditions and norms of behaviour
on human responses in addition to the impact of the immediate physical
environment. This is particularly true in naturally ventilated buildings. This
complexity implies that an adaptable (or a variable) standard is needed which
takes all these factors into account, and one that includes the effects of season
and climate,allows buildings to change, suggests how quickly they should do
so, and reflects the willingness of occupants to vary their environment by
giving some measure of control to them. In this way the development of a
standard methodology would allow more useful comparisons to be made
between various studies carried out in different countries around the world.

Application difficulties do not affect the validity of thermal comfort research, but question its utility. In recent years a number of difficulties have arisen when applying heat exchange equations to conditions encountered in every day living for the season already described.

Since in daily life fluctuations are commonplace, because of changes in clothing, in activity and in local conditions, the general application of steady-state equations becomes inappropriate and their use will produce not only random errors but also impose a systematic bias. Furthermore, human thermal sensations do not settle down at a constant level, even in a steady environments, but oscillate with a period of several minutes (Sakura,et al,1991). Recent results from climate chambers shows that the assumption that comfort temperature for standard metabolic activity and standard clothing is not always true. Of particular interest, are the recently obtained experimental values from climate chamber studies of Malay students in London and Malaysia [2]. They show a highly significant 3 °C difference in comfort temperature.

At any judgement level in a given environment there will be a particular percentage of people dissatisfied; the same people in an other space could make different percentage judgements even if the thermal environment is similar, because of other factors being different. [9] shows that as extremes of heat or cold are approached there will be a convergence of judgements (see Fig. 11).

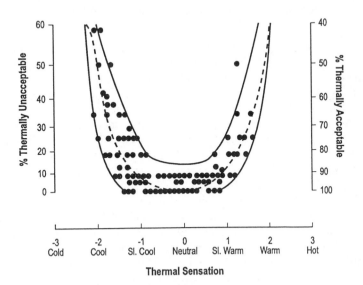

Figure 11: Mean thermal acceptability and thermal sensation votes (Berglund, 1979)

DJ Clements-Croome

Experiments on rapid changes in temperature show an asymmetrical response in the thermal sensation (Rohles, et al, 1980). People are more sensitive to cold than heat and this will be depicted as an asymmetrical distribution of votes reflecting the thermoreceptor sensitivity curves for the skin described by Geldard [49] and Zhao [145], (see Fig. 12).

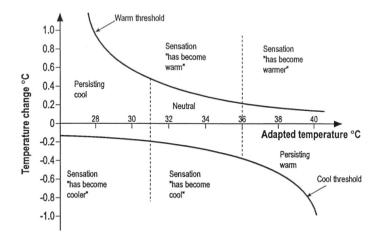

Figure 12: Psychophysics of the temperature sense

Again it is unlikely that the thermal judgement curves will be linear although it is always assumed that they are. In experiments on clothing, Kim and Tokura [65] states that the subjects dressed faster with thicker clothing in the morning than in the evening, and the subjects with face cooling dressed faster with thicker clothing than the subjects without face cooling.

Current comfort standards prescribe a static "ideal" temperature that is to be maintained uniformly over space and over time. This is unrealistic and can lead to wasteful fuel consumption. The experience of an environment at any moment is dependant on ones past experiences and so a time sequence analysis is important as people adapt physiologically and behaviourally in accordance with changes, expectations and preferences.

There is growing dissatisfaction with the static comfort temperatures predicted using traditional models based on heat balance theory and laboratory data. This is coupled with concern about the increasing amounts of energy required for maintaining the thermal environment in buildings and its attendant environmental impacts. Static indoor temperature standards encourage the use of high-energy environmental control strategies,and exclude options for which temperature variation is either inevitable or desirable, (e.g. many passive, energy-conserving solutions, or innovative mechanical environmental control strategies). In comparison, an adaptable (or a variable) temperature standard that recommends temperatures which reflect the climate surrounding the building would reduce the indoor-outdoor temperature differential and could be expected to reduce energy requirements considerably (Auliciems 1990).

Humphreys [55,53] has collated data from numerous field studies made in several countries which demonstrate that most of the variation in the indoor temperature required for comfort can be explained by the changes in the monthly mean out door temperature (see Fig. 13).

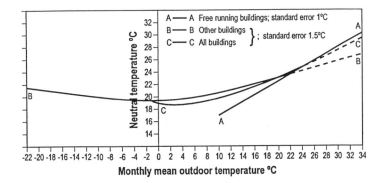

Figure 13: Neutral temperature curves (Humphreys, 1995)

For the case of free-running buildings, there is a strong linear correlation between monthly mean outdoor temperature and the indoor comfort temperature; recent work by Nicol [92] at al suggest the exponentially weighted running mean outdoor temperature may improve the correlation further. Auliciems and de Dear [93] analysed a large number of field surveys, excluding some of Humphreys data but including some more recent studies, and derived another comfort temperature equation based on mean outdoor temperature.

The adaptive approach recognises that people use numerous strategies to achieve thermal comfort. They are not inert recipients of the environment, but interact with it to optimise their conditions. The adaptive, people-centred way of regarding thermal comfort suggests that it would be advantageous to re-formulate temperature standards for buildings, so that they reflect the empirical relation between climate and thermal comfort and make due allowances for human adaptability.

The ASHRAE Standard 55 has a rudimentary allowance for climate, in that it advocates temperatures which differ between summer and winter. Adaptive results can be used to refine this allowance by linking the indoor comfort temperature to the outdoor temperature throughout its seasonal and geographical variation. This would result in increased design flexibility without reducing user satisfaction. It could also lead to reducing the capacity of installed heating and cooling plant, and thus save energy.

For the case of free-running buildings, there is a strong linear correlation between monthly mean outdoor temperature and the indoor comfort temperature, whereas for other buildings there is a fairly strong curvilinear relationship which can be improved if the mean daily maximum temperature of the hottest month is used as an additional predictor. Statistical analysis of the field data showed that for free-running buildings (Humphreys,1970):

$$\theta_n = 11.9 + 0.534 * \theta_o$$

Where θ_n is the predicted neutral temperature for thermal comfort and θ_o is the mean outdoor temperature for the months being considered; this regression equation has a coefficient of correlation equal to 0.97 and range of application is: $10\ ^\circ C < \theta_o < 33\ ^\circ C$.

The indoor environment is affected by building construction as well as the services systems. Buildings have an infiltration characteristic; well insulated buildings usually have higher internal surface temperatures and are warmer. Massive buildings are usually cooler than lighter ones. Warm moist atmospheres can directly contribute to sick building syndrome but also indirectly, because they can encourage microbial growth, and thus affect concentrations of air pollutants. Air movement has an effect on thermal comfort, but it may have an independent effect on some symptoms because it effects heat, moisture and pollutant distribution [30]. Indoor environment is a dynamic combination of physical, chemical, social, and biological factors, which in total affect human health, well-being and comfort.

Wyon [139] states that thermal comfort should be regarded as a by-product of the interaction between task and environmental effects on the behaviour of the subject. Thermal comfort is only one of several considerations regarding the choice of thermal conditions indoor. It is concluded that moderate heat stress, only a few degrees centigrade above the optimum, has a marked effect on mental performance when temperatures rise slowly. Tasks demanding concentration and clear thinking are adversely affected, but memory and cue utilization can be improved by temperatures up to 26 °C, declining rapidly thereafter in the case of male subjects. In hot weather, work involving concentration should therefore be carried out early in the day. Memory and tasks requiring an increased depth of attention can, with advantage, be postponed until the temperature has risen, provided that it does not rise above 26°C. Woman are generally more sensitive to temperature conditions than men [139].

The group average neutral temperature for thermal comfort is of little interest for the individual as 95% of individual preferred neutral temperatures can vary by more than 10 °C, even under standard conditions (Wyon, 1993). Provision of individual control for as many environmental factors as possible, makes it possible to eliminate the discrepancy between individual and group responses and thus provide a cost effective means of increasing productivity. Individual differences with regard to the heat balance requirements, as well as thermal preferences, emphasise the need for increased degrees of freedom of individual control for air quality and thermal factors. This is also true of other environmental factors.

6 Healthy buildings

Indoor environment can be defined by physical features of the environment
such as lighting, colour, temperature, air quality and noise. These factors have
been studied extensively with regard to their impact on task performance and
satisfaction. Sundstrom [124] reports laboratory studies that show with high
consistency that ambient conditions do have real and meaningful influences on
behaviour in 150 out of a total of 185 experiments. A complete analysis of
indoor environmental quality would take into consideration not only indoor air
quality and thermal comfort but also lighting, floor lay-out, colour scheme,
building materials, noise level, disruption, weather, management styles, space,
employee and customer backgrounds, employee/customer satisfaction and
employee motivating factors. Any field study must account for all of these
factors.

Interior finishes can have a marked effect on air quality. Lisbet [74] shows
that occupants working in offices with carpets complained significantly more
about the sensation of dry and stuffy air than occupants working in offices
with linoleum flooring, Research by Rudnai (1994) (See Fig. 14) shows the
prevalence of the most frequent symptoms among people living in buildings
with different types of construction in Budapest. Brick construction shows a
lower incidence of SBS than concrete because the latter buildings were warmer
and drier, due to poor heating control and possibly low infiltration. The mixed
concrete and breeze block homes suffered from higher concentrations of
formaldehyde. Clearly, material and construction are important.

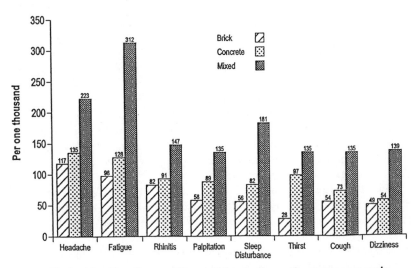

**Figure 14: Prevalance of the most frequent symptoms among people
living in different types of building in Budapest**

Ricci-Bitti [105] distinguish between *sick building syndrome* (SBS) and *mass psychogenic illness* (MPI). The former is a temporary failure to cope with the environment whereas the latter represents a collective stress response. *Building related illness* (BRI) and *Neurotoxic disorders* (NTD) are deemed to be caused by environmental pollution, although minor forms of the these syndromes may be related to stress. The principal difference between these phenomena is that neurotoxic disorders and building related illness (such as legionnaires disease) persists whether the person is in the workplace or away from it, but in the case of sick building syndrome and mass psychogenic illness the effects only occur in the working environment and tend to decay quickly on leaving it. Most attention has been given to the sick building syndrome.

The World Health Organisation (WHO, 1983) describe the term sick building syndrome as a collection of non-specific symptoms expressing a general malaise associated with occupancy of the workplace. Symptoms include irritations of the eyes, respiratory system as well as mental discomforts due to lethargy or headaches. The ancient Chinese believed that invisible lines of energy, known as *chi* run through our bodies and the environment. A smooth flow of chi means wellbeing, but if it is blocked then ill health, expressed in various ways, results. *Feng Shui* is the art of freeing and circulating chi. Equipment such as computers emit electro-magnetic fields and these together with underground water and geological faults can disturb the earths natural electro-magnetic field. There remains the possibility that geopathic stress could be partially responsible for sick building syndrome.

There have been numerous studies on sick building syndrome, but no co-ordinated research and became the methodology of investigation has varied it is difficult to compare results. Recently a study by Jones [63], together with the work on standard questionnaires by Raw [103] means future investigations can be carried out in a more consistent manner. It is, for example, very difficult to make any strong conclusions about the effect of ventilation with respect to sick building syndrome. According to work by Jaakkola [62] and also Sterling (1983) ventilating a building with 25% or 100% outdoor air makes little difference. Likewise other work referred to by Hedge [50] quotes work where the ventilation rate has been increased from 10 l/s per person to 25 l/s per person with no beneficial effects on SBS symptoms; Sundell [122] and also found that ventilation rates above 10 l/s per person have no discernable effect on the symptoms.

Jones [63] classifies healthy offices as those which average 1-2 symptoms per worker, while relatively unhealthy offices have 4-5 symptoms per worker. Headaches and lethargy appear to be nearly always the most frequently reported symptoms.

Hedge [50] carried out an investigation in six office buildings and found that the prevalence of eye, nose and throat symptoms were higher in airconditioned offices than naturally ventilated ones, but headaches did not show any consistent pattern. Other studies by Burge et al [13]; Mendell & Smith [87]; Wilson & Hedge [134] confirmed that SBS symptoms were less prevalent in naturally ventilated buildings than airconditioned ones, but some mechanically ventilated buildings did not give rise to any problems either. Robertson et al [106] compared SBS symptoms in adjacent air-conditioned and naturally ventilated offices and found that the symptoms were more prevalent among workers in the airconditioned offices, although measurement of a variety of physical environmental factors failed to show any significant differences in the environmental conditions between the buildings. Other work described by Hedge [50] confirms these findings. However, care should be taken to ensure that the concentrations of volatile organic compounds are measured and compared. Some studies have found that symptoms maybe associated with suspended particulate matter, but there is contrary evidence about this issue also. The lack of a consistent association between symptoms and the physical environment suggests that building sickness syndrome remains elusive and so maybe there are a number of other factors like geopathic stress, individual factors, perceived control and occupational factors are more important. Again psychosocial factors are probably not directly related to the SBS symptoms, but Hedge [50] argues that psychosocial variables may trigger off patterns of symptom reporting. Hedge [50] describes a large study that he has undertaken in 27 airconditioned offices. Dry eyes and headaches were found to be weakly associated with formaldehyde; mental fatigue was found to weakly associated with formaldehyde and particulate concentrations. Complaints about stale air were associated with carbon dioxide levels. The building users age, job grade and smoking status was not associated with SBS symptoms. Results showed that more SBS were reported by women; full-time computer users; building occupants with high job stress or low job satisfaction; people who perceived the indoor air quality to be poor; the 18-35 years old age group; occupants who have allergies; people who have migraines; users who wore spectacles or contact lenses, and finally smokers. Hedge [50] concluded that reports about sick building syndrome symptoms are influenced by several individual and occupational interacting factors. Other studies by Hedge [50] attempted to establish if a certain personality type was susceptible to sick building syndrome but did not establish any connection. Likewise the effect of circadian rhythms did not appear to be important. Sick buildings syndrome appears to a rise from a set of multiple risk factors some of which are environmental, but biological, perceptual and occupational aspects are also important. The general conclusion can be made that environmental, occupational and psychosocial factors interact to induce sick building syndrome symptoms.

Jones [63] considers health and comfort in offices and states that the symptoms commonly referred to as sick building syndrome (SBS) show a wide range of variation in reported symptoms between buildings and also between zones in any one building. Only about 30% of the variation in symptoms has been explained by the built environment, by an individuals characteristics or by the job.

In common with other work he concludes that the symptoms of SBS arise from a combination of factors such as building environment, nature of the work being undertaken, as well as the individual characteristics of the person. One important conclusion reached by Jones [83] is that when thermal conditions are perceived as comfortable, the office is not necessarily healthy. This is also in some ways analogous to earlier work, already discussed, which suggests that the most productive environments are slightly less than comfortable.   He calls for improved standards of maintenance, spatial simplicity and flexibility, as well as adaptability of services, good management and recognition that the building, its environment and activities all interact dynamically.

The Health and Safety Executive published a booklet in 1995 entitled *How to Deal with Sick Building Syndrome*. This booklet helps building owners to identify and investigate buildings, besides giving advice on how to create a good work environment.   As regards productivity, it is pointed out that although the SBS symptoms are often mild they do not appear to cause any lasting damage. They can affect attitudes to work and can result in reduced staff efficiency; increased absenteeism; staff turnover; extended breaks and reduced overtime; lost time due to complaining. Advice is offered in the HSE Booklet on building services and indoor environment; finance; and job factors, including management systems and work organisation.

Whitely et al [129] considers reasons why it has been difficult to explain the range of individual differences in symptom reporting in the same building and between buildings.   Equity theory suggests that the way workers are treated may affect productivity. If this is true then a person in an organisation may intuitively assess their level of reward from the organisation and put the amount of effort in which is related to this.   The physical environment in which a person works, can be seen as part of the reward system.   The organisation can influence peoples attitudes depending on whether any attempt is made to improve conditions in the workplace.

Whitely [130] describes a well established, relatively stable, personality measure which is related to perceived control called locus of control, which measure the general tendency to attribute outcomes of behaviour due to internal or external causes, and goes on to conclude that the locus of control and job satisfaction appear to explain peoples perceptions of the environment as a whole.   These factors also significantly influence the way people report sick building syndrome, environmental conditions and productivity.   Whitely [130] state that self reports of productivity change due to the physical environment must be treated with extreme caution, as job satisfaction is a major factor also. Research has already been described which shows that the perception of control over the environment is important with regard to productivity.

7 Indoor environment and productivity

Traditionally thermal comfort has been emphasized as being necessary in buildings but is comfort compatible with health and well-being? The mind and body need to be in a state of health and well-being for work and concentration, which are a prime prerequisite for productivity. Good productivity brings a sense of achievement for the individual as well as increased profits for the work organization. The holistic nature of our existence has been neglected because knowledge acquisition by the classical scientific method has dominated research and is controlled but limited, whereas the world of reality is uncontrolled, subjective and anecdotal but nevertheless is vitally important if we are to understand systems behaviour. It is possible to reconsider comfort in terms of the quality of the indoor environment and employee productivity.

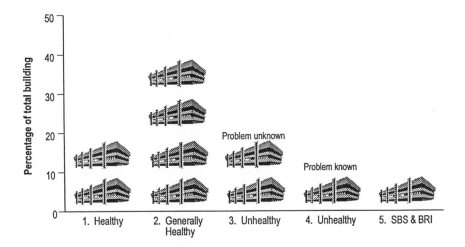

Figure 15: Buildings by health category (Dorgan, 1994)

Dorgan [37] has analysed some 50,000 offices in the USA. Fig. 15 shows the percentage of existing commercial buildings in that are healthy. The description of each category is as follows: (i) Healthy Buildings: Always meet ASHRAE Standards 62-1989 and 55-1992 during occupied periods; (ii) Generally Healthy Buildings: meet ASHRAE Standards 62-1989 and 55-1992 during most occupied periods; (iii) Unhealthy Building: fail to meet ASHRAE Standards 62-1989 and 55-1992 during most occupied period (iv) Buildings with Positive SBS: more than 20% of occupants complain of more than two SBS symptoms, and frequently 6 of the more common 18 SBS symptoms.

Health is the outcome of a complex interaction between the physiological, personal and organisational resources available to the individual and the stress placed upon them by their physical environment, work, and home life,

Symptoms occur when the stress on a person exceeds their ability to cope and where resources and stress both vary with time so that it is difficult to predict outcomes from *single causes*. Sickness building syndrome is more likely with warmer room conditions and this can will lead to decreased productivity. Higher temperatures also mean wasteful energy consumption. When temperature reaches uncomfortable levels, output is reduced. On the other hand, output improves when high temperatures are reduced by air conditioning. When temperatures are either too high or too low, error rates and accident rates increase. While most people maintain high productivity for a short time under adverse environmental conditions, there is a temperature threshold beyond which productivity rapidly decrease [75]. Mackworth [79] stated that overall the average number of errors made per subject per hour increased at higher temperatures and showed that the average number of mistake per subject per hour under the various conditions of heat and high humidity was increased at higher temperatures especially above 32°C.

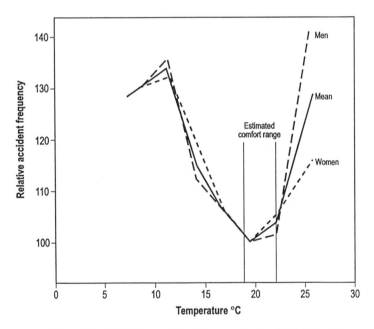

Figure 16:   Relative accident frequencies for British Munitions
plant workers at different temperature (Vernon, 1936)

Fig. 16. demonstrates relative accident frequencies for British munitions plant workers at different temperatures; the accident frequency was a minimum at about 20°C in three munition factories [128].

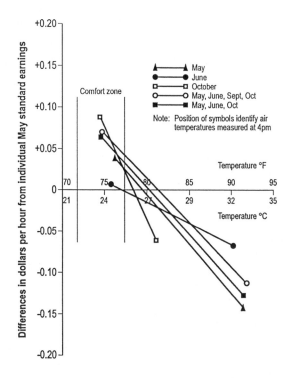

Figure 17: **Regressions of individual productivity on indoor air temperature adjusted for differences in monthly group productivity (Pepler, 1963)**

Fig. 17. shows regressions of individual productivity on indoor air temperature adjusted for differences in monthly group productivity. It clearly shows that variations in productivity in a non- airconditioned mill were influenced by temperature changes although absenteeism was apparently not related to the thermal conditioning; on average an 8% productivity increase occurred with a 5°C decrease in temperature (Pepler, 1963).

Factors, such as poor lighting, both natural and artificial, poorly maintained or designed air conditioning, and poor spatial layouts are all likely to affect performance at work. This may be evidenced by lower performance. In a survey of 480 UK offices occupiers, Richard Ellis states that 96% were convinced that the design of a building affects productivity and when asked an open ended question to categorise the aspects of design that they felt would tend to lead to this effect; 43% used words such as attractive, good visual stimulus, colours and windows; 41% mentioned good morale, 'feel good' factor and contented happy staff; 19% said more comfortable, relaxing, restful conditions to work; 16% said increases in motivation and productivity; 15% said improved communications; 3% or less said reduce stress. All these aspects help to promote a well designed building. The importance of various factors are summarised in the following table and it can be seen that natural daylight and ventilation are rated highly, but green issues and the use of atria are also significant.

Feature	Very	Quite	Not-very	Not-At-all
best use of natural daylight	57%	31%	10%	1%
ventilation using windows	30%	41%	25%	3%
thermal design for building	12%	40%	36%	6%
energy-saving green design	15%	36%	37%	8%
use of atria & glazed streets	4%	20%	52%	18%

Table 5. Importance ratings of various designs factors (Ellis 1994)

Clearly any building that does not maximise its natural day-lighting is likely to be unpopular with office occupiers. The high value attributed to the use of windows rather than airconditioning partly reflects the generally low level of effectiveness achieved by airconditioning in many buildings, but also more fundamentally the inherent need for natural light and good views out of the building.

Wilkins [131] reports that good lighting design practice, particularly the use of daylight, can improve health without compromising efficiency. Concerns about the detrimental effects of uneven spectral power distribution and low-frequency magnetic fields are not as yet substantiated. Wilkins [131] states that several aspects of lighting may affect health, including (i) low-frequency magnetic fields; (ii) ultra-violet emissions; (iii) glare; and (iv) variation in luminous intensity. The effects of low-frequency magnetic fields on human health are uncertain. The epidemiological evidence of a possible contribution to certain cancers cannot now be ignored, but neither can it be regarded as conclusive. The ultraviolet light from daylight exceeds that from most sources of artificial light. Its role in diseases of the eye is controversial, but its effects on skin have been relatively well documented. The luminous intensity of a light source, the angle it subtends at the eye, and its position in the observer's visual field combine to determine the extent to which the source will induce a sensation of discomfort or impair vision. Glare can occur from the use of some the lower intensity sources, such as the small, low-voltage, tungsten-halogen lamps. It is reasonable to suppose that in the long-term, glare can have secondary effects on health and that is visibly flickering can have profound effects on the human nervous system. At frequencies below about 60 Hz it can trigger epileptic seizures in those who are susceptible. In others it can cause headaches and eye-strain. Wilkins [131] concludes that the trend to towards brighter high-efficiency sources is unlikely to affect health adversely, and may indeed be advantageous. The trend could have negative consequences for health were it to be shown that the increasing levels of ambient light at night affect circadian rhythms. Improvements in brightness and the evenness of spectral power may be beneficial. In particular, the move towards a greater use of daylight is likely to be good for both health and efficiency.

In many buildings, users report most dissatisfaction with temperature and ventilation, while noise, lighting and smoking feature less strongly. The causes lie in the way temperature and ventilation can be affected by changes at all levels in the building hierarchy, and, most fundamentally, by changes to the shell and services. In comparison, noise, lighting and smoking are affected mainly by changes to internal lay out and work station arrangements which can often be partly controlled by users.

There are some indications that giving occupants greater local control over their environmental conditions improves their work performance and their work commitment and morale which all have positive implications for improving overall productivity within an organization. Building users are demanding more control at their work stations of fresh air, natural light, noise and smoke. Lack of control can be significantly related to the prevalence of ill health symptoms in the office environment, and there is widespread agreement that providing more individual control is beneficial. Work by Burge et al [13] (see Fig. 18) demonstrates the relationship between self-reports of productivity and levels of control over temperature, ventilation and lighting. The graph shows that the productivity increases as individual control rises across all the variables.

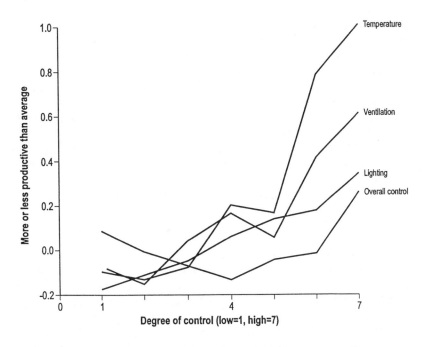

Figure 18: The relationship between self-reports of productivity and levels of control over temperature, ventilation, lighting and overall control

Intervention to ensure a healthy working environment should always be the first step towards improving productivity. There are very large individual differences in the tolerance of sub-optimal thermal and environmental conditions. Even if the average level of a given environmental parameter is appropriate for the average worker, large decrements in productivity may still be taking place among the least tolerant. Environmental changes which permit more individual adjustment will reduce this problem. Productivity is probably reduced more when large numbers work at reduced efficiency than when a few hypersensitive individuals are on sick leave. Wyon [139] states that commonly occurring thermal conditions, within the 80% thermal comfort zone, can reduce key aspects of human efficiency such as reading, thinking logically and performing arithmetic, by 5-15%.

Lorsch and Abdou [1] summarise the results of a survey undertaken for industry on the impact of the building indoor environment on occupant productivity, particularly with respect to temperature and indoor air quality. They also describe three large studies of office worker productivity with respect to environmental measurements, and discuss the relationship between productivity and building costs.

It is felt in general that improving the work environment increases productivity. Any quantitative proof of this statement is sparse and controversial. There are a number of interacting factors which affect productivity, including privacy, communications, social relationships, office system organisation, management, as well as environmental issues. It is a much higher cost to employ people who work than it is to maintain and operate the building, hence spending money on improving the work environment may be the most cost effective way of improving productivity. In other words if more money is spent on design, construction and maintenance and even if this results in only small decreased absenteeism rates or increased concentration in the workplace, then the increase in investment is highly cost effective [137].

In one case study reported by Lorsch and Abdou [178] it is not clear if the drop in productivity was due to a reduction in comfort, by the loss of individual control or frustration due to being inconvenienced. According to Pepler and Warner (1968) people work best when they are slightly cool, but perhaps not sufficiently so to be termed discomfort, and should not to be cool for too long.

According to a report by the National Electrical Manufacturers Association in Washington [90] increased productivity occurs when people can perform tasks more accurately and quickly over a long period of time. It also means people can learn more effectively and work more creatively, and hence sustain stress more effectively. Ability to work together harmoniously, or cope with unforeseen circumstances, all point towards people feeling healthy, having a sense of well-being, high morale and being able to accept more responsibility. In general people will respond to work situations more positively. At an ASHRAE Workshop on *Indoor Quality* held at Baltimore in September 1992 the following productivity measures were recommended as being significant.

- Absence from work or work station.
- Health costs including sick leave, accidents and injuries.
- Interruptions to work.
- Controlled independent judgements of work quality.
- Self assessments of productivity.
- Speed and accuracy of work.
- Output from pre-existing work groups.
- Cost for the product or service.
- Exchanging output in response to graded reward.
- Volunteer overtime.
- Cycle time from initiation to completion of process.
- Multiple measures at all organisational levels.
- Visual measures of performance, health and well-being at work.
- Development of measures and patterns of change over time.

Rosenfeld [109] describes that when airconditioning was first conceived it was expected that the initial cost of the system would be recovered by an increased volume of business. He quotes an example where the initial cost of the airconditioning system for an office building is about £100 per m², so that if the average salary is £3,000 per m² and there is an occupancy of 10 m² per person, then adding 10% to the cost of the system is justified if it increases productivity by as little as 0.33%. Such small differences are difficult to measure in practice. Figure 19 shows the relationship between the savings in working hours and the incremental initial cost of the system for a range of salaries.

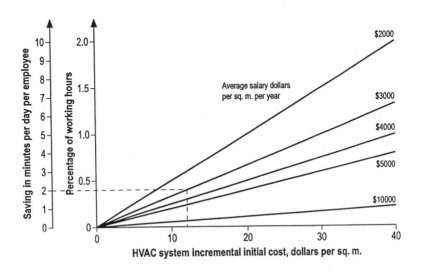

Figure 19: The relationship of incremental initial cost
to potential user time saving (Rosenfield, 1989)

Rosenfeld [109] shows that improvement in indoor air quality can be more than offset by modest increases in productivity. This leads to the conclusion that in general, high quality systems which will have higher capital costs can generate a high rate of return in terms of productivity. In addition systems will be efficient, be effective, have low energy consumptions and consequently achieve healthier working environments in buildings with a low CO_2 emission.

Holcomb and Pedelty [53] attempt to quantify the costs of potential savings that may accrue by improving the ventilation system. The increase in cost can be offset by the gain in productivity resulting from an increase in employee work time. Higher ventilation rates generally result in improved indoor air quality. Collins [23] reported that 50.1% of all acute health conditions were caused by respiratory conditions due to poor air quality. Cyfracki [32] reported that a productivity increase of 0.125% would be sufficient to offset the costs of improved ventilation. It should be mentioned again that some studies have shown a decrease in SBS symptoms with increased ventilation rates [113] while others have not [89]. [53] conclude that although there is some inconsistency there is still sufficient evidence to suggest that there is an association between ventilation rates, indoor air quality, sick building syndrome symptoms and employee productivity.

Lorsch and Abdou [78] conclude that temperatures which provide optimum comfort may not necessarily give rise to maximum efficiency in terms of work output. The difficulty here is that this may be true for relatively short periods of time, but if a person is feeling uncomfortable over a long period of time it may lead to a decrement in work performance. However, there is a need for more research in this area. It almost seems that for optimum work performance a keen sharp environment is needed which fluctuates between comfort and slight cool discomfort.

It is easier to assess the effects of temperature on physical performance, but much more difficult to test the effects on mental performance. For example, the lowest industrial accident rate occurs around 20°C and rises significantly above or below this temperature. The other problem is the interaction with other factors which contribute towards the productivity. Motivated workers can sustain high levels of productivity even under adverse environmental conditions for a length of time which will depend on the individual.

Lorsch and Abdou [77] analyse several independent surveys which show that when office workers find the work space environment comfortable, productivity tends to increase when airconditioning is introduced by as much as 5-15%, in the opinion of some managers and researchers. These are however only general trends and there is little hard data and some findings are contradictory. Kobirck and Fine [67] conclude that it is difficult to predict the capabilities of groups of people, never mind individuals, in performing different tasks under given sets of climatic conditions.

A study for the Westinghouse Furniture Systems Company in Buffalo, New York, entitled *The Impact of the Office Environment on Productivity and the Quality of Working Life* [11] suggested that the physical environment for office work may count for a 5-15% variation in employee's productivity. And the general conclusion was that people would do more work on an average work day if they are physically comfortable.

Woods [136] reported that satisfaction and productivity vary with the type of heating, ventilation or airconditioning system. Central systems appeared to be more satisfactory than local ones, the most important factor being whether there was cooling or not. In one study on user controlled environmental systems by Drake [38] the ability to have local control was important in maintaining or improving job satisfaction, work performance and group productivity, while reducing distractions from work. For example, some users reported that they wasted less time taking informal breaks compared to times when environmental conditions were uncomfortable. They were also able to concentrate more intensively on their work. The gain in group productivity from the user controlled environmental system amounted to 9%. A number of studies suggest that a small degree of discomfort is acceptable, but it has to be confined to a point where it does not become a distraction.

Work by Kamon [64] and others shows that heat can cause lethargy which can increase the rate of accidents, and also affect productivity. Bedford [9] concluded that there was a close relationship between the external temperature and the output of workers. Deteriorating performance is partially contributable to insomnia due to heat. Schweisheimer [114] carried out some surveys concerned with establishing the effect of airconditioning on productivity at a leather factory in Massachusetts, an electrical manufacturing company in Chicago and a manufacturing company in Pennsylvania. In all cases after the installation of airconditioning the production increased by between 3-8.5% during the Summer months. On the basis of these investigations Schweisheimer [115] concluded that the average performance of workers dropped by 10% at an internal room temperature of 30°C, by 22% at 32°C and by 38% at 35°C.

Konz and Gupta [67] investigated the effects of local cooling of the head on mental performance in hot working environments. The subject had to create words in ten minutes from one of two sets of 8 letters, which were type printed on a blank form. Poor conditions without cooling resulted in the creation of words dropping by some 20% in the hot condition, where as with cooling the reduction was only some 12%.

Abdou and Lorsch [1] studied the effects of indoor air quality on productivity. It was concluded that productivity in the office environment is sensitive to conditions leading to poor indoor air quality and this is linked to sick building syndrome. It is recognised that any stress is also influenced by management and other factors in the workplace. Occupants having local control over their environment generally have an improvement in their work effort, but in a more general way there is a synergistic effect of a multitude of factors which effect the physical and mental performance of people. Abdou and Lorsch [1] conclude that *in many case studies occupants have been highly dissatisfied with their environment, even though measurements have indicated that current standards were being met.* This highlights the need to review standards and the basis on which they are made. Exactly the same conclusion is made by Donnini et al [36].

Although, it is difficult to collect hard data which would give a precise relationship between the various individual environmental factors and productivity there is sufficient evidence to show that improved environment decreases peoples complaints and absenteeism, thus indirectly enhancing productivity. The assessment of problems at the work place, using complaints is unreliable, because there is little mention of issues that are working well, and also the complaints may be attributable to other entirely different factors. Abdou and Lorsch [1] contend that the productivity of 20% of the office work force in the USA could be increased simply, by improving the air quality of offices, and this is worth approximately $60 billion per year.

Work by Vernon [126,127] shows that there is a clear relationship between absenteeism and the average ventilation grading for a space, which was judged by the amount of windows on various walls, so that windows on 3 sides had the highest grading and windows on one side only had the lowest. Abdou and Lorsch [1] give the following causes as being the principal ones contributing to sick building syndrome:

- Building occupancy higher than intended.
- Low efficiency of ventilation.
- Renovation using the wrong materials.
- Low level of facilities management.
- Condensation or water leakage.
- Low morale and lack of recognition.

In this case lower efficiency of ventilation means that the supply air is not reaching the space where the occupants are, hence the nose is breathing in recirculated stale air. It is important to realise that even if the design criteria are correct for ventilation, the complete design team are responsible for ensuring that the systems can be easily maintained; the owner and the facilities manager also need to ensure that maintenance is carried out effectively. The tenant and occupants should use the building as intended. When new pollutant sources are introduced, such as new materials or a higher occupancy density, then the ventilation will become inadequate.

Burge [14] conducted a study of building sickness among 4373 office workers in 42 office buildings having 47 different ventilation conditions in the United Kingdom. The data was further analysed by Raw [103]. The principal conclusions were that as individuals reported more than two symptoms, the subjects reported a decrease in productivity; none of the best buildings in this survey were airconditioned and these had fewer than two symptoms per worker on average, whereas the best airconditioned buildings had between two and three symptoms; women recorded more symptoms than men, but there was no overall difference in productivity; individual control of the environment has a positive effect on productivity; the productivity is increased by perceived air quality; productivity, however, only increases with perceived humidity up to a certain point and then appears to decrease again. Evidence supporting the importance of individual control of environment is again provided by Preller [102]. It should be said that some contrary evidence exists concerning some of these factors, which points to the need for a systems approach to studying the effects of environment in buildings such as that proposed by Jones [63].

Productivity can be related to quality and satisfaction of the service or functional performance. Studies have shown that productivity at work bears a close relationship to the work environment. Burge [14] demonstrates that there is a strong relationship between self-reports of productivity and ill health symptoms related to buildings: productivity decreases as ill health symptoms increase. There is a slightly less marked trend relating productivity and air quality but there is a significant effect.

Dorgan [37] defines productivity as the increased functional and organizational output including quality. This increase can be the result of direct measurable decreases in absenteeism, decreases in employees leaving work early; or reductions of extra long breaks and lunches. The increase can also be the result of an increase in the quantity and quality of production while employees were active; improved indoor air quality is an important consideration in this respect. There is general agreement that improved working conditions, and the office environment is certainly one of the more important working conditions, tend to increase productivity. However, determining a quantitative relationship between environment and productivity proves to be highly controversial. While some researchers claimed they reliably measured improvements of 10% or more, others present data showing that no such relationships exist. Since the cost of the people in an office is an order-of magnitude higher than the cost of maintaining and operating the building, spending money on improving the work-environment may be the most cost-effective way to improve worker productivity [75].

In 1994, the energy use in an average commercial office building in the United States costs approximately $20/m^2/year, whereas the functional cost is approximately $3,000/m^2/year. The functional cost includes the salaries of employees, the retail sales in a store, or the equivalent production value of a hotel, hospital, or school. This means that 1% gain in productivity ($30/m^2/year) has a larger economic benefit than a 100% reduction ($20/m^2/year)in energy usage. In addition, the productivity gains will increase the benefits such as repeat business in hotels, faster recovery times in hospitals, and attainment of better jobs due to a better education in schools. A small gain in worker productivity has major economic impacts and it makes sense to invest in improving the indoor environment to achieve productivity benefits. Dorgan [37] states that productivity gains of 1.5% in generally healthy buildings, and 6% in sick buildings, can be easily achieved. As typical pay-back costs of improvement in the indoor air ranges from less than 9 months to 2 years, the benefits clearly offset the renewal cost resulting in a very favourable cost-to-benefit ratio. The 1.5% improvement is conservative. Some literature indicates that this may be as high as 5 to 10%. However, achieving such productivity gains may require using advanced active or passive environmental control as well as personal controls. Examples of productivity gains in the order of 1 - 3 % are found in several studies [68]. Informal (unpublished) and anecdotal reports on productivity gains have been performed in supermarkets, fast food outlets, retail department store, schools, and office buildings resulting in estimated gains in sales ranging from 4 to 15% in retail stores during summer months [37].

By focusing on the productivity benefits, projects which improve the indoor environment are moved away from an energy-saving viewpoint and more towards a productivity-increase viewpoint. Even if a proposed project improves the indoor environment but increases the energy cost by 5%, the project may still be economically feasible if the productivity increase is greater than 0.04%. Wyon [139] states that the "leverage" of environmental improvements on productivity is such that a 50% increase in energy costs of improved ventilation would be paid for by a gain of only 0.25-0.5% in productivity, and capital investments of $50/m^2 would be paid for by a gain of only 0.5% in productivity. The pay-back time for improved ventilation is estimated to be as low as 1.6 years on average, and to be well under 1 year for buildings with ventilation that is below currently recommended standards.

An increase in productivity can be achieved with either (i) no increase in energy usage or even a decrease (ii) with an increase in funding for the given level of technology. The use of energy recovery systems, and the increased use of such technologies as advanced filtration, dehumidification, thermal storage and natural energy are all examples of energy improving technologies the cost of which can be offset by increase in productivity. The increased building services budget can allow for the introduction of the best system, not the cheapest. Any indoor environment productivity management program should be able to include reducing energy consumption as one of the design objectives. Improving indoor environment will provide a high return on investment through productivity gains, health saving and reduced energy use. The benefits of improved indoor environment are improved productivity, increased profits, greater employee-customer-visitors health and satisfaction, and reduced health costs. The potential productivity benefits of improved indoor environment are so large that this opportunity cannot be ignored. There are indirect, long-term, and social benefits.

8 Assessment and measurement of productivity

The economic assessment of environment both in terms of health (medical treatment, hospitalization) and of decrease in productivity (absenteeism) has had very little attention by researchers as yet. However this assessment is absolutely necessary in order to assess the effectiveness of improved design and management protocols [6]. Since there are no standard procedures to measure productivity, it has been difficult to get professionals to accept the concept of a relationship between economic productivity benefits and indoor environment. The challenge is to develop a method to implement the link between indoor environment and productivity using scientific principles whilst taking into account the experience of occupational psychologists for example.

Thorndik [125] described four criteria for performance measures: validity, reliability, freedom from bias, and practicality.The inclusion of validity and reliability in the list of criteria implies that the standards for performance measures are similar to those of the tests. Freedom from bias most frequently originates from two relatively independent mechanisms. The first of these is that of rater bias. In this case, those who use the rating may systematically rate the performance of particular individuals either higher or lower across a number of dimensions than is justifiable from the rated performance. Finally, the concern for practicality as a measurement criterion is obvious. Ironically, in spite of its obvious importance, very little attention is given to the discussion of issues of practicality in the literature. There has been little disagreement about the four criteria for performance measurement over the last 40 years.

Ilgen [57] states that the methods of performance measurement can be classified into three categories: (i) *Physiological*; (ii) *Objective*; and (iii) *Subjective*. The rationale for using *physiological methods* is based on the reasoning that physiological measures of activation or arousal are associated with increased activity in the nervous system which is equated with an increase in the stress on the operator. Common physiological measurements include: (a) cardiovascular measures: heart rate, blood pressure; (b) respiratory system: respiration rate, oxygen consumption; (c) nervous system: brain activity, muscle tension, pupil size; and (d) biochemistry: catecholamine. Three fundamental criticisms have been raised against physiological measures of workload by Meister [85]. First, he questions the validity of the measures. The evidence supporting a relationship between physiological and other workload indices is not strong, and the meaning of these relationships when they do occur is frequently difficult to interpret. Secondly, the measures themselves are highly sensitive to contaminating conditions. Thirdly, he argues that the measures are intrusive and/or impractical. Physiological measures can restrict or interfere with the operator's task performance, and restrictions imposed by the job (e.g. task demands, safety considerations) can often limit the number and kind of physiological measures used at one time.

The second of three classes of measures of mental workload has been labelled *objective measures* [90]. Task performance is frequently used to infer the amount of workload, both mental and physical. Task measures are typically divided into primary and secondary task measures. In primary task performance measurement, task difficulty for a single task is manipulated and performance variations are assumed to reflect change in workload. In secondary or comparative task performance measurement, the person is first presented with a single task and then a second task is added, or performance is compared across two different tasks and changes in performance are recorded. Task-based measurement has advantages in that it has high face validity and it is amenable to quantitative/empirical testing. Task measures present a number of challenges. However, first conclusions based on the task performance allude to the limited resource model, namely that individuals have a finite pool of resources which can be devoted to one task or distributed among tasks. If this model does not hold, then the conclusions from this method are not valid. Secondly, this procedure has a great deal of utility in the laboratory where task performance and the introduction of new tasks can be highly controlled [85]. Thirdly, it is often difficult to cross-calibrate (scale) diverse measures across tasks.

A final set of measures of workload is comprised of *subjective measures* [89]. Subjective measures of workload are applied to gain access to the subjects' perceptions of the level of load they are facing in task performance. Rating scales, questionnaires, and interviews are used to collect opinions about workload. While these methods may not have the empirical or quantitative appeal of physiological or objective measures, it is often argued that subjective measures are most appropriate since individuals are likely to work in accordance with their feelings regardless of what physiological or behaviourial performance measures suggest [89]. An assumption of subjective assessment is that people are ware of, and can introspectively evaluate changes in their workload, and that this assumption holds regarding general impressions of the difficulty of ongoing experiences. When comparing subjective measures with performance measures, high correlations have been found during early and middle stages of overload [78]. Usually higher subjective ratings of workload correspond to poorer performance, yet there is evidence that respondents rate workload higher in a task that they perform better. The advantages of subjective measures as a group include ease of implementation, non-intrusiveness, low cost, face validity, sensitivity to workload variations, and the wide variety of available techniques [94].

There is some evidence that air pollution, or the perception that it exists, can create stress among employees who believe that it poses a threat to their health. The stress may be particularly intense among people who believe they have no control over the pollution. It is further argued that psychosocial factors, such as labour-management relations, and satisfaction or dissatisfaction with other factors in the work environment, can have a profound influence on the level of response of the occupants to their environment. Productivity measurement may be carried out by using physiological methods, which has been described already, and environmental psychological methods.

Early studies (Mehrabian and Russell, 1974) showed that the emotional and behaviour effects due to environment may be more easily assessed than physiological responses. Mehrabian (1974) states that there are three basic emotional responses (pleasure, arousal, and dominance) which combine and can be used to describe any emotional state. The effects of the physical environment on emotional and behaviourial responses to work performance can be compared by considering the impacts of environment on pleasure, arousal and dominance. Environmental psychology has been concerned with two major topics: the emotional impact of physical stimuli and the effect of physical stimuli on a variety of behaviours such as work performance or social interaction.

Cooper [24] states that if you were to take a look at stress levels in the organisation that you would find people reporting a negative attitude toward the indoor environment would also be the people expressing high job dissatisfaction or low mental well-being. Generally speaking, occupational stress is regarded as a response to situations and circumstances that place special demands on an individual with negative results. Occupational stress is regarded as a response to situations and circumstances that place special demands on an individual with negative results. Cooper [24] designed a occupational stress indicator to gather information on groups of individuals. The indicator has six questionnaires entitled: how one feels about the job; how one assesses their current state of health; the way one behaves generally; how one interprets events around them; sources of job pressure; and, how one copes with general stress. The sources of stress are multiple, as are the effects. It is not just a function of being "under pressure". The sources may be work-related, but home life will also be implicated. The effects in terms of health may not just concern how one feels physically but how one reacts and behaves in work and at home. The Occupational Stress Indicator (OSI) attempts to measure: (i) the major sources of occupational pressure; (ii) the major consequences of occupational stress and (iii) coping mechanisms and individual difference variables which may moderate the impact of stress (Cooper, 1988). An environmental dimension can be built into this indicator(see Fig. 20 and Fig. 21). Questionnaires and semi-structured interviews are used for subjective assessment. A study of the medical records and the attendance of the staff members will also be made.

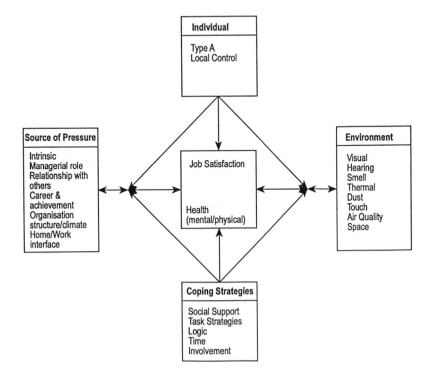

Figure 20: Conceptual basis for impact occupational stress indicator

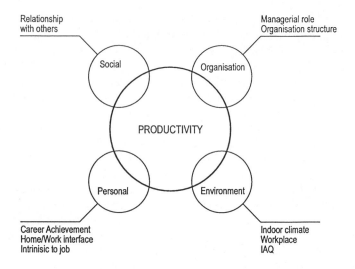

Figure 21: Conceptual basis for impact environment on productivity

A number of research studies have demonstrated that there is no simple relationship between single environmental factors and complex human behaviour. The Analytic Hierarchy Process (AHP) method together with multi-regression and correlation analysis can be used to establish an empirical model of "multi-sensory" well-being of occupants. AHP is an effective process to quantitatively describe unquantitative systems and factors [111]. The system should first be divided into various factors according to their properties and the main objective. These factors are then classified in successively subordinate grades or levels through which the analytic hierarchical model can be developed. The importance of the factors in the lowest hierarchy can be related to the factors in the highest hierarchy on the basis of questionnaire and semi-structured interview surveys [143]. On the basis of analysing factors such as the visual, hearing, smell, thermal, dust, touch, fresh, and space (see Fig. 22), it is to hoped to develop a relationship between "multi-sensory" occupant well-being (indoor environment) and productivity.

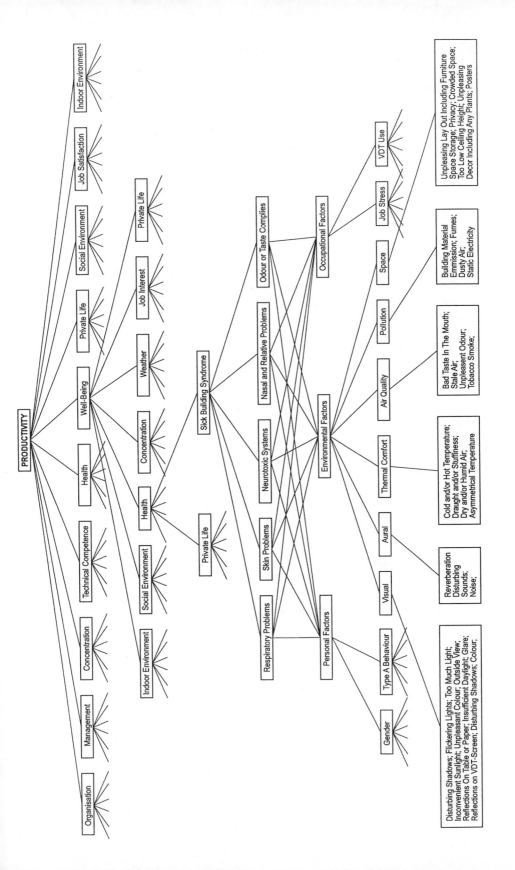

Figure 22: Priority factors influencing productivity and well-being of occupants for use in AHP model

9 Conclusions

In summary, this Chapter proposes that comfort should be viewed in the context of well-being and hence link the quality of the indoor environment with employee productivity. The published information about well-being, health, and productivity benefits has been reviewed. Assessment of productivity has been discussed. It is proposed that research is needed in the following areas:

- The meaning of comfort and the differences between the terms comfortable, acceptable, preferable and tolerable thermal environments.
- The link between productivity, well-being and comfort or discomfort.
- The relationship between thermal comfort and other design requirements such as air quality, noise and light.
- Assessment of the Analytic Hierarchy Process for studying the interaction of environment with productivity thus establishing the priority factors for the design process.
- Optimal design and the relationship between temperature, economics, health, productivity, energy use and comfort within the overall context of the management and production processes.

There is a need to develop an empirical model to enable greater understanding of "multi-sensory" well-being of occupants under "realistic" dynamic working conditions and to develop a correlation between "multi-sensory" occupant comfort, well-being and productivity. Most work concentrates on thermal comfort for groups of people and this is unrealistic for a basis of environmental design. This alternative holistic approach will enable standards to be evolved which are realistic and recognise the combined value of low energy, health, comfort and productivity in various situations.

10 Acknowledgement

Li Baizhan carried out part of the literature survey and is developing the AHP work as part of his doctoral studies.

11 References

1. Abdou, O. A., Lorsch, H. G., (1994 c.) The Impact of the Building Indoor Environment on Occupant Productivity: Part 3: *Effects of Indoor Air Quality*, ASHRAE Trans. 1, 100(2), pp. 902-913.

2. Abdulshukor, (1993) *Human Thermal Comfort in the Tropical Climate*, PhD thesis, University of London, UK.

3. Aizlewood, C. E. et al. (1994) *The New Comfort Equation: Where Are We Now?*, Building Research Establishment, PD123/94, June, BRE/109/1/46, pp. 1-8.

4. ASHRAE, (1989) *ASHRAE Standard 62-1989: Ventilation for Acceptable Indoor Air Quality*, ASHRAE Atlanta, USA.

5. ASHRAE, (1992) *ASHRAE Standard 55-92: Thermal Environmental Conditions for Human Occupancy*, ASHRAE Atlanta, USA.

6. Barbatano, L. et al. (1994) *Home Indoor Pollution. A Contribution for an Economic Assessment*, Proceedings of Healthy Air'94, Italy, pp. 395-403.

7. Bauer, R. M., Grevu, K. W., Besh, E. L., Schruarke, C. J., Hicks, A., Ware, M. R. and Lyles W. B. (1992). The role of psychological factors in the report of building-related symptoms in sick building syndrome, *Journal of Consulting and Clinical Psychology*, 50, pp. 213-219.

8. Bedford, T. (1949) *Air Conditioning and the Health of the Industrial Worker. Journal of the Institute of Heating and Ventilating Engineers* 17(166): pp. 112-146.

9. Berglund, L. G. (1979) *Thermal Acceptability*, ASHRAE Transactions, ASHRAE Atlanta, USA, pp. 825-834.

10. BOSTI. (1982) *The Impact of the Office Environment on Productivity and the Quality of Working Life*. Buffalo, NY: Westinghouse Furniture Systems.

11. Brager, G. S. (1992) *Using Laboratory Based Models to Predict Comfort in Offices*, ASHRAE J., Vol. 34, No. 4, pp. 44-49.

12. Brager, G. S. et al. (1994) *Comparison Of Methods For Assessing Thermal Sensation and Acceptability in the Field*, Proceedings of Thermal Comfort: Past, Present and Future, Garston, BRE.

13. Burge, S. A. et al. (1987) *Sick Building Syndrome: A Study of 4373 Office Workers*, Ann.Occ.Hygiene, 31, pp. 493-504.

14. Burge, S., Hedge, A., Wilson, S., Harris-Bas, J. and Robertson, A. (1987) Sick building syndrome: A study of 4373 office workers. *Annals of Occupational Hygiene* 31: pp. 493-504.

15. Building Research Establishment Digest (226), (1979) *Thermal, Visual And Acoustic Requirements In Buildings*, June, pp. 1-7.

16. CEN/TC 156/WG 6 Doc N 100, DRAFT ENV XXX, (1994) *Ventilation for Buildings Design Criteria for the Indoor Environment*, April, pp. 1-59.

17. Cena, K, M. et al. (1990) *A Practical Approach to Thermal Comfort Surveys in Homes and Offices: Discussion of Methods and Concerns*, ASHRAE Transactions, USA, Vol.96, Part 1, pp. 853-857.

18. Chen Qigao, et al. (1994) *Report On Indoor Thermal Environment Test*, Chongqing Jianzhu University, China.

19. CIBSE, (1986) *CIBSE Guide*, Volume A Design Data, London, Chartered Institution of Building Services Engineers.

20. Clark, R. P. and Edholm, O. G. (1985) *Man And His Thermal Environment*, Edward Arnold.

21. Cohen, S. (1980) *Cognitive Processes As Determinants of Environmental Stress:, Stress and Anxiety*, Vol. 7, New York: Hemisphere, pp. 171-184.

22. Collins, J. G. (1989) *Health characteristics by occupation and industry: United States 1983-1985*. Vital Health and Statistics 10(170). Hyattsville, MD: National Center for Health Statistics.

23. Commission of the European Communities, Directorate General for Science, Research and Development, Joint Research Centre -- Environment Institute, European concerted action *Indoor Air Quality & Its Impact on Man*, Environment and Quality of Life, report No. 10 *Effect of Indoor Air Pollution on Human Health*.

24. Cooper, C. L. and Robertson. (1990) *International Review Of Industrial And Organizational Psychology*, John Wiley & Sons Ltd, Vol.9.

25. Crombie, A. C. (1995) *Styles of Thinking in the European Tradition*, Duckworth & Co.

26. Croome, D. J. (1981) *Airconditioning and Ventilation of Buildings*, Second Edition, Pergamon Press, pp. 1-575.

27. Croome, D. J. (1991) *The Determinants of Architectural Form in Modern Building Within the Arab World*, Building and Environment, Vol. 26, No. 4, pp. 349-362.

28. Croome, D. J. et al. (1992) *Evaluation Of Thermal Comfort And Indoor Air Quality In Offices*, Building Research and Information, Vol.20, pp. 211-225.

29. Croome, D.J., (1980), *Man, Environment And Buildings*, Building and Environment, Vol. 15, pp. 235-238.

30. Croome, D. J. and Rollason, D. H. (1988) *Healthy Buildings* No. 88, Vol. 2, Stockholm, Sweden, pp. 393-402.

31. Cyfracki, L. (1990) *Could upscale ventilation benefit building occupants and owners alike?* Proceedings of Indoor Air '90: 5th International Conference on Indoor Air Quality and Climate, Vol. 5, pp. 135-141. Aurora, ON: Inglewood Printing Plus.

32. David, A. (1994) *The Role Of The Building Services Engineer*, Building Services, July, pp. 36-38.

33. de Dear, R. (1993) *Outdoor Climatic Influences On Indoor Thermal Comfort Requirements*, Thermal Comfort: Past, Present and Future, ed., N. Oseland, Building Research Establishment, Watford, United Kingdom.

34. de Dear, R., and Fountain, M. (1994) *Field Experiments On Occupant Comfort And Office Building Thermal Environments In A Hot-Humid Climate*. ASHRAE Transactions, Vol. 92, (2).

35. Dick, J. B., (1971), Building International, March-April, 93.

36. Donnini, G. et al. (1994), *Office Thermal Environments and Occupant Perception of Comfort*, La Riforma Medica, Vol. 109, Supp 1, (2), pp. 257-263.

37. Dorgan, C.E. et al. (1994) *Productivity Link to the Indoor Environment Estimated Relative to ASHRAE 62-1989*, Proceedings of Health Buildings'94, Budapest, pp. 461-472.

38. Drake, P. (1990) *Summary of Findings from the Advanced Office Design Impact Assessments.* Report to Johnson Controls, Inc. Milwaukee, WI.

39. Drake, P., Mill, P. and Demeter, M. (1991) *Implications of user-based environmental control systems: Three case studies.* ASHRAE Transactions 98(2).

40. Ellis, R. (1994) *Tomorrow's Workplace*, A Survey For Richard Ellis by The Harris Research Centre.

41. Eto, J. H. and C. Meyer. (1988) *The HVAC Costs of Fresh Air Ventilation.* ASHRAE Journal vol. 30 No. 9 pp. 31-35.

42. Fanger, P. O. et al. (1986) *Perception of Draught in Ventilated Spaces.* Ergonomics. London, England: Taylor and Francis Ltd, Vol. 29, No. 2, pp. 215-235.

43. Fanger, P. O. et al. (1989) *Air Turbulence And Sensation Of Draught*, Energy and Buildings, Lausanne, Switzerland: Elsevier Sequoia S.A, Vol. 12, pp. 21-39.

44. Fanger, P. O. and Christensen, N. K. (1987) *Prediction Of Draft*, ASHRAE Journal, January, pp. 30-31.

45. Fanger, P. O. (1988) *Introduction of the Olf and the Disposal Units to Quantify Air Pollution Perceived By Humans Indoors and Outdoors*, Energy and Buildings, No. 12, pp. 1-6.

46. Fanger, P. O. (1970) *Thermal Comfort*, McGraw-Hill, New York.

47. Fanger, P. O. (1982) *Thermal Comfort*, Robert E. Krieger Pub. Co., Mala bar, Florida.

48. Gagge, A. et al. (1986) *A Standard Predictive Index of Human Response to the Thermal Environment*, ASHRAE Trans. 92, Pt.2.

49. Geldard, F. A. (1972) *The Human Senses*, John Wiley & Sons, Inc. New York.

50. Hedge, A. (1994) *Sick Building Syndrome: Is It An Environment Or A Psychological Phenomenon?* La Riforma Medica, Vol. 109, Supp. 1, (2), 9-21

51. Hesketh, B. (1993) *Measurement Issues In Industrial And Organizational Psychology*, International Review of Industrial and Organizational Psychology, Vol. 8, Edited by Cooper C.L, John Wiley & Sons Ltd, No. 4, pp. 133-172.

52. Holcomb, L. C., Pedelty, J. F., (1994) *Comparison of Employee Upper Respiratory Absenteeism Costs with Costs Associated with Improved Ventilation*, ASHRAE Trans., 100 (2), 914-920

53. Humphreys, M. A. (1995) Chapter 1 in *Standards for Thermal Comfort* edited by Nicol et al, E F Spon.

54.   Humphreys, M. A. (1993) *Field Studies and Climate Chamber Experiments in Thermal Comfort Research*, Building Research Establishment paper 52/72, June, pp. 52-69.

55.   Humphreys, M.A. (1976) *Field Studies Of Thermal Comfort Compared and Applied*, Buildings Services Engineer, Vol. 44.

56.   Humphreys, M.A. (1970) J. Inst. Heat. Vent. Engrs., Vol. 38, pp. 95.

57.   Ilgen, D. R. and Schneider, J. (1991) *Performance Measurement: A Multi-discipline View*, International Review of Industrial and Organizational Psychology, 1991, Vol. 6, Edited by Cooper C.L. and Robertson, John Wiley & Sons Ltd, Chapter 3, pp. 71-108.

58.   ISO, (1984) *Standard 7730: Moderate Thermal Environments -- Determination of the PMV and PPD Indices and Specification of the Conditions for Thermal Comfort*, Geneva, Switzerland: International Standards Organization.

59.   ISO, (1985) *ISO-7726: Thermal Environments -- Instruments and Methods for Measuring Physical Quantities*, Geneva: International Standards Organization.

60.   Istvan, L. (1994) *The physiological, Psychological and Social Relation Between Man and His Environment*, Proceedings of the 3rd International Conference: Healthy Buildings'94, Vol. 1, pp. 71-76.

61.   Jaakkola, J. K. J., Heinonnen, O. P. and Seppänen, O. (1989), *Sick Building Syndrome, Sensations of dryness and thermal comfort in relation to room temperature in an office building: need for individual control of temperature*, Environmental International, Vol. 15, pp. 163-168

62.   Jaakkola, J. K. J., Heinonen, O. P. and Seppänen, O. (1991) *Mechanical Ventilation in Office Building Syndrome.* An Experimental Epidemiological Study. Indoor Air 2, pp. 111-121.

63. · Jones, P. (1995) *Health and Comforting Offices*, The Architects Journal, June 8th, 33-36.

64.   Kamon, E. (1978) *Physiological and Behaviourial Responses to the Stresses of Desert Climate.* In Urban Planning for Arid Zones, G. Golany, ed. New York: John Wiley and Sons.

65.   Kim, H. E. and Tokura, H. (1994) *Effects of Time of Day on Dressing Behaviour Under the Influence Of Ambient Temperature Fall From 30°C to 15°C*, Physiology & Behaviour, Vol. 55, No. 4, pp. 645-650.

66.   Kobrick, J. L. and B. J. Fine. (1983) *Climate and Human Performance.* In the Physical Environment at Work, D.J. Oborne and M.M. Gruneberg, eds., pp. 69-197. New York: John Wiley and Sons.

67.   Konz, S. and Gupta, V. K. (1969) *Water-Cooled Hood Affects Creative Productivity.* ASHRAE Journal 40:43.

68.   Kroner, W. M. and Stark-Martin, J. A. (1992) *Environmentally Responsive Work Stations and Office Worker Productivity*, Presented at Workshop on Environment and Productivity, June.

69. Latham, G. P. et al. (1993) *The Increasing Importance if Performance Appraisals to Employee Effectiveness in Organizational Settings in North America*, International Review of Industrial and Organizational Psychology, Vol. 8, Edited by Cooper C.L. and Robertson, John Wiley & Sons Ltd, Chapter 3, pp. 87-132.

70. Levin, H. (1991) *IAQ, Productivity, and Occupant Control*, Indoor air Bulletin, December, Vol. 1, No. 7.

71. Li, B. and Croome, D. J. (1995) *The Prediction of Indoor Climate for Occupant Cooling in Ventilated Building*, TSINGHUA-HVAC-'95, 2nd International Symposium on HVAC, China.

72. Li, B. (1990) *Thermal And Environmental Design to Meet Future Needs*, Proceedings of Tall Building 2000 and Beyond Conference, Hong Kong, Vol. 2, pp. 365-374.

73. Li, B. (1992) *Investigation of Thermal Environment of House in China's Southern Part and Research of Ways to Solve Existing Problem*, International Journal for Housing Science and Its Application, USA, Vol.16, No. 2, pp. 115-122.

74. Lisbet, B. and Gunnar L. (1994) *Indoor Climate Complaints And Symptoms of the Sick Building Syndrome in Offices*, Proceedings of Health Buildings'94, Budapest, pp. 391-396.

75. Lorsch, H. G. and Ossama, A. A. (1993) *The Impact of the Building Indoor Environment on Occupant Productivity*, ASHRAE Transactions, Vol. 99, Part 2, USA.

76. Lorsch, H. G. and Abdou, O. A., (1994 a.) *The Impact of the Building Indoor Environment on Occupant Productivity: Part I: Recent Studies, Measures and Costs*, ASHRAE Trans., 100 (2), 741-749.

77. Lorsch, H. G. and Abdou, O. A., (1994 b.) *The Impact of the Building Indoor Environment on Occupant Productivity: Part II: Effects of Temperature*, ASHRAE Trans. 100(2), pp. 895-901.

78. Lysaught, R. J. et al. (1989) *Operator Workload: Comprehensive Review and Evaluation of Operator Workload Methodologies* (Analytic Tech. Rep. 2075-3), Alexandria, VA: US Army Research Institute for Behavioral and Social Sciences.

79. Mackworth, N. H. (1946) *Effects of Heat on Wireless Operators Hearing and Recording Morse Code Messages*, Bri. J. Ind. Med. 3: pp. 143-158.

80. Markus, T. et al. (1972) *Building Performance by Building Performance Research Unit*, Strathclyde University, Applied Science.

81. Marshall, A. H. (1967) PhD Thesis, Institute of Sound and Vibration Research, University of Southampton.

82. McBride, M. F. *ASHRAE Standard 90.2P building envelope requirements*, pp. 725-746.

83. McCartney, K. J. and Croome, D. J. (1994) *A Study of the Effects f User-Controlled Localised Environmental Systems on Comfort, Energy and Productivity*, Proceedings of the 3rd International Conference: Healthy Buildings'94, Vol. 1, pp. 273-279.

84. McNall, P. E. (1979) ASHRAE Trans., 85, Part I.

85   Meister, D. (1986) *Human Factors: Testing And Evaluation*, Amsterdam: Elsevier.
86.   Melius, J., Wallingford, K., Keenlyside, R. and Carpenter, J. (1984) *Indoor Air Quality - The NIOSH Experience*. National Symposium of the American Institute of Architects, San Fransico, CA.
87.   Mendell, M. J., and Smith, A. H. (1990) *Consistent Pattern of Elevated Symptoms in Airconditioned Office Buildings*, 80, pp. 1193-1199.
88.   Menzies, R., Tamblyn, R., Farant, J., Hanley, R. T., Tamblyn, F., Nunes, F. and Marcotte, P. (1991) *The Effect of Varying Levels of Outdoor Ventilation on Symptoms of Sick Building Syndrome*. IAQ '91: Healthy Buildings, pp. 90-96. Atlanta: American Society of Heating, Refrigerating and Airconditioning Engineers, Inc.
89.   Moray, N., et al. (1979) *Final Report of the Experimental Psychology Group*, Mental Workload: Its Theory and Measurement, edited by Moray N., New York: Plenum, pp. 101-11463.
90.   NEMA (1989), *Lighting and Human Performance*: A review. A report sponsored by the Lighting Equipment Division of the National Electrical Manufacturers Association, Washington, DC, and the Lighting Research Institute, New York.
91.   Nicol, F., Humphreys, M. et al. (1995) *Standards For Thermal Comfort*, E8 & FN SPON London, UK.
92.   Nicol, J. F. and Humphreys, M. A. (1973) *Thermal Comfort as Part of a Self-Regulating System, Building Research and Practice*, May/June, pp. 174-179.
93.   Nicol, F. et al. (1995) EPSRC Grant GR/J 77214.
94.   O'Donnell, R. D. and Eggemeier, F. T. (1986) *Workload Assessment Methodology*, Handbook of Perception and Human Performance: Cognitive Processes and human performance, edited by Boff, K.R., et al. New York: Wiley.
95.   Olesen, B. W. and Kjerulf-Jensen, P. *Energy Consumption in a Room Heated by Different Methods*, p19-29.
96.   Ornstein, S. (1990) *Linking Environmental and Industrial/Organizational Psychology*, International Review of Industrial and Organizational Psychology, 1990, Vol. 5, Edited by Cooper C.L. and Robertson, John Wiley & Sons Ltd, Chapter 7, pp. 195-228.
97.   Oseland, N. A. (1993) *A Summary of the Proceedings of a Conference Entitled: Thermal Comfort: Past, Present and Future.*
98.   Oseland, N. A. (1994) *A Within Groups Comparison of Predicted and Reported Thermal Sensation Votes in Climate Chambers, Offices and Homes*, Proceedings of the 3rd International Conference: Healthy Buildings '94", Vol. 1, pp. 41-47.
99.   Oseland, N. A. (1994) *A Review of Thermal Comfort and its Relevance to Future Design Models and Guidance*, Proceedings of BEPAC Conference -- Building Environmental Performance Facing the Future, April, UK, pp. 205-216.
100.   Oseland, N. A. and Humphreys, M. A. (1994) *Trends in Thermal Comfort Research*, BRE Report 266, BRE, Garston, Watford, UK.

101. Perler, R. D. (1963) *Temperature: Its Measurement and Control in Science and Industry*, Chapter 3, (Ed. Hardy) Reinhold.

102. Preller, L., Zweers, T., Brunekreef, B. and Boleij, J. S. M. (1990) Indoor Air '90, *Fifth International Conference on Indoor Air Quality and Climate* 1: pp. 227-230.

103. Raw, G. J., Roys, M. S. and Leaman, A. (1990) *Further Findings From the Office Environment Survey: Productivity.* Indoor Air Quality '90, Fifth International Conference on Indoor Air Quality and Climate 1: pp. 231-236.

104. Rees, Sir Dai. (1995) *Science for the Goodlife*, Times Higher Education Supplement, August 11th, Page 11.

105. Ricci-Bitti, P. E., Caterina, R. (1994), La Riforma Medica, Vol. 109, Supp 1, (2), pp. 215-223.

106. Robertson, A. S., Burge, P. S., Hedge, A., Sims, J., Gill, F. S., Finnegan, M., Pickering, C. A. C. and Dalton, G. (1985), Comparison of health problems related to work and environment measurements in two office buildings with different ventilation systems. *British Medical Journal*, 291, 373-376.

107. Rohles, F. H. and Nevins, R. G. (1974) *The Nature of Thermal Comfort for Sedentary Man*, ASHRAE Transactions, Vol. 77 (1), pp. 239-246.

108. Romeo, C. and Trisciuoglio, A. (1994) *Indoor Air Quality with Specific Reference to School Rooms. Methods of Assessment and Diagnosis*, Proceedings of Healthy Air'94, Italy, pp. 57-62.

109. Rosenfeld, S. (1989) *Worker productivity: Hidden HVAC Cost.* Heating/Piping/Air Conditioning, September pp. 69-70.

110. Ryan, C. M. and Morrow, L. A. (1992) *Dysfunctional Buildings and Dysfunctional People; an Examination of the Sick Building Syndrome and Allied Disorders.* Journal of Consulting and Clinical Psychology, 60, PP. 220-224.

111. Saaty, T. L. (1972) *The Analytic Hierarchy Process*, McGraw-Hill, New York, USA.

112. Sakura Y., et al. *Transient Response of Human Sensory System to Changes in Thermal Environment*, ASHRAE Transactions, Vol.97 (2), pp. 170-178.

113. Samimi, B. S. and Seltzer, J. M. (1992) *Sick building syndrome due to insufficient and/or nonuniform fresh air supply in a multi storey office building.* IAQ '92: Environments for People, pp 319-322. Atlanta American Society of Heating, Refrigerating and Airconditioning Engineers, Inc.

114. Schweisheimer, W. (1962) *Does Air Conditioning Increase Productivity?* Heating and Ventilating Engineer 35(419): PP. 669.

115. Schweisheimer, W. (1966) *Erhaehung und Leistung und Produktion*, Warme Luftungs und Gesundheitstechnik, Nov., 278-265.

116. Shirom, A. (1986) *On The Cross-environmental Generality of the Relational View of Stress*, Journal of Environmental Psychology, Vol.6, pp. 121-134.

117. Shove, E., (1995) *Personal Communication with Centre for the Study of Environmental Change at Lancaster University.*

118. Sterling, E. M. and Sterling, T. (1983) The impact of different ventilation levels and fluorescent lighting types on building illness: an experimental study. *Canadian Journal of Public Health,* 74, pp. 385-392.

119. Sterling, E. M. et al. (1993) *Building Design, Technology, and Occupant Well-Being on Temperate Climates,* Proceedings of International Conference on Building Design, Technology, and Occupant Well-being on Temperate Climates, Belgium, pp. 1-407.

120. Suess, M. J. (1994) *Indoor Air Problems And Quality,* Proceedings of the 3rd International Conference: Healthy Buildings'94", Vol. 1, pp. 17-26.

121. Sundell, J., Lindvall, T. and Stenberg, B. (1991) *Influence of type of ventilation and outdoor airflow rate on the prevalence of SBS symptoms.* IAQ '91: Healthy Buildings, pp. 85-89. Atlanta: American Society of Heating, Refrigerating and Airconditioning Engineers, Inc.

122. Sundell, J. (1994) On the Association Between Building Ventilation Characteristics, Some Indoor Environmental Exposures, Some Allergic Manifestations, and subjective symptom reports. *Indoor* Air, Supp. 2.

123. Sundstrom, E. (1986) *Work Place: The Psychology of the Physical Environment in Offices and Factories,* New York: Cambridge University Press.

124. Sundstrom, E. (1987) *Work Environments: Offices and Factories,* Handbook of Environmental Psychology", Vol. 1, New York: pp. 733-783.

125. Thorndike, R. L. (1949) *Personnel Testing,* New York: Wiley.

126. Vernon, H. M., Bedford, T. and Warner, C. G. (1926) *A Physiological Study of the Ventilation and Heating in Certain Factories.* Rep. Industry. Fatigue Res. Bd., No. 35. London.

127. Vernon, H. M., Bedford, T. and Warner, C. G. (1930) *A Study of Heating and Ventilation in Schools.* Rep. Industry. Health Res. Bd., No. 35. London.

128. Vernon, H. M. (1936) *Accidents and Their Presentation,* Cambridge University Press.

129. Whitley, T. D. R., Makin, P. J., Dickson, D. J. (1995a) *Organisation and Job Factors in Sick Building Syndrome,* Proceedings of Healthy Buildings '95, Milan, 11-14 Sept.

130. Whitley, T. D. R., Makin, P. J., Dickson, D. J. (1995b) *The Environment, Comfort and Productivity,* Proceedings of Healthy Buildings '95, Milan, 11-14 Sept.

131. Wilkins, A. J. (1993) *Health And Efficiency in Lighting Practice,* Energy, Vol. 18, No. 2, pp. 123-129.

132. Wilkins, A. J. et al. (1990) *Inferences Regarding the Visual Precipitation Of Seizures, Eye-strain and Headaches,* Generalised Epilepsy: Neurobiological Approaches, Birkhauser, Boston, USA, pp. 314-326.

133. Wilson, M. Dubiel, J. Nicol, F. (1995) *Thermal and Acoustic Comfort* Proc. SAVE Conference, Passive Cooling of Buildings, Athens.

134. Wilson, S. and Hedge, A. (1987) *The Office Environment Survey: A Study of Building Sickness*, Building Use Studies Ltd., London, UK.

135. Woods, J. E. et al. (1981) *Relationships Between Measures of Thermal Environment and Measures of Worker Productivity*, ASHRAE Transactions, Part 1, pp. 117-144.

136. Woods, J. E., Drewry, G. M. and Morey, P. (1987) *Office Worker Perceptions of Indoor Air Quality Effects on Discomfort and Performance.* Fourth International Conference on Indoor Air and Climate, pp. 464-468, Berlin, Germany.

137. Woods, J. E. (1989) Cost avoidance and productivity in owning and operating buildings. In Occupational Medicine, State of the Art Reviews 4(4), Cone, J. E. and Hodgson, M. J., eds. and *Problem Buildings: Buildings Associated Illness and the Sick Building Syndrome*, pp. 753-770. Philadelphia: Hanley & Belfus.

138. World Health Organisation, (1983), *Indoor Air Pollutants Exposure and Health Effects*, EURO Reports and Studies 78, World Health Organisation.

139. Wyon, D. P. (1993) *The Economic Benefits of a Healthy Indoor Environment*, Proceedings of Heathy Air'94, Italy, pp. 405-416.

140. Wyon, D. P. (1991) *The Ergonomics of Healthy Buildings: Overcoming Barriers to Productivity*, Supplementary Proceedings of IAQ 91 Conference, ASHRAE, USA, pp. 43-36.

141. Wyon, D. P. (1986) *How does Indoor Climate Affect Productivity and Performance?* VVS & Energy, No. 3, pp. 59-65.

142. Wyon, D. P. et al. (1982) *The Effect of Moderate Thermal Stress on the Potential Work Performance of Factory Worker*, Energy and Buildings, April.

143. Yao Runming, Liu Antian, and Li Baizhan, (1992) *Application of the AHP Method to the Planning of a Solar House in China*, Proceedings of Thermal Performance of the Exterior Envelopes of Buildings V, December, Clearwater Beach , Florida, USA, pp. 602-606.

144. Zhang Guo Gao, (1989) *Physiology and Human Behaviour at Higher Temperatures*, Press House of Science and Technology in Shanghai, China, pp. 1-251.

145. Zhao, R., et al., (1995), *Thermal Acceptability in Transient Environment*, Proceedings of 2nd International Symposium on Heating, Ventilating and Airconditioning, Sept. 25-27, Beijing, 74-80.

CHAPTER FOUR

The User's Role in Environmental Control: Some Reflections on Theory in Practice

Dean Hawkes
Welsh School of Architecture, University of Wales,
Cardiff UK

Abstract
A significant element of the development of the theory and practice of environmental design in buildings in recent years has been the move away from wholesale reliance upon automatic controls of plant and other systems and the restoration of control to the occupants. In work first published in the 1980s the author proposed a distinction between two *modes* of environmental control the *exclusive* and the *selective*. The selective mode was characterised by the exploitation of ambient energy sources and the use of occupant control. This theoretical proposition was first applied in practice in the design and construction of a primary school at Netley in Hampshire, designed by the Hampshire County Architect's Department. Recently the principles of selective design have become widely adopted in the design of buildings for a range of uses. This paper offers a critical reconsideration of the theory in the light of this practical experience and proposes further guidelines for the effective control of building environments by their users.

Naturally Ventilated Buildings: Buildings for the senses, the economy and society. Edited by D. Clements-Croome. Published in 1997 by E & FN Spon. ISBN 0 419 21520 4

1 The theory of selective design

The theory of 'Selective'design [2] rests upon the idea of a building as a system of interacting elements (Fig 1).

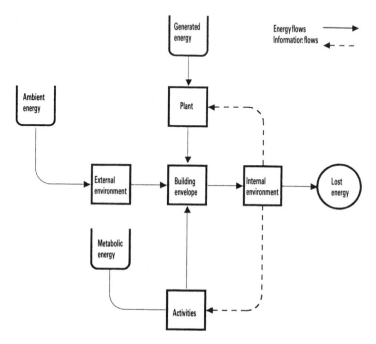

Fig 1   The Environmental Control System

These comprise the fabric of the building, its form, the mechanical service systems which it may have, and their control mechanisms. These are located within a given climate which is, itself, described in some detail to capture the variations which occur, seasonally and diurnally, in solar radiation, temperature, humidity, wind speed and direction, levels of natural light and so forth. It is also important to note the impact on the local environment of a building which follows from human actions. These include the effects of adjoining buildings in overshadowing and obstructing the site, but should also acknowledge noise sources and atmospheric pollution. Finally the 'Selective' system explicitly defines the needs and the role of the occupants of a building in its environmental processes, bringing us full circle to the question of comfort, and to a further dimension to its definition, the active participation of occupants in defining and controlling the environment.

A number of years ago a research study was undertaken in a group of school buildings in Essex as part of a research project carried out at the Martin Centre in the Department of Architecture at Cambridge. In this four fundamental questions were posed concerning the internal environment of buildings [3].

1. What is the potential for variation of environmental conditions which is offered by buildings of different types?
2. To what extent do the occupants of buildings take active steps to modify the environment, and at what point?
3. How wide is the range of conditions which is tolerated?
4. Does this "toleration" demand changes in activity patterns?

The studies showed that great emphasis was placed by the occupants of these buildings, in this case the teachers, upon the ability to control the environment of their space. Buildings with automatically-controlled mechanical plant caused the greatest dissatisfaction. From this it was inferred that the ability to exercise some control over the environment in a building may have a crucial bearing upon psychological satisfaction. It was also found that people manipulated the controls which their buildings offered in a clear response to the state of the external climate. For example, they would adjust the condition of the building, by, for example, opening windows or adjusting solar control blinds, as the climate demanded. They would also often anticipate the effects of climatic conditions before they were manifested in the building by exercising a kind of 'feed-forward' response. Blinds would be closed before the sun struck the windows, for example. It was also discovered that teachers would manipulate the environment of a classroomsorrow to create an appropriate ambience for a particular lesson, such as vigorous, well-ventilated conditions for teaching arithmetic, or a cosy, womb-like state for story-telling to young children. This phenomenon may be relatively unique to the environmental design of certain building types, for example such contrivance may be inappropriate in an office building or a factory, but it serves to reveal a level of environmental sophistication which exceeds that implied by orthodox theory.

The work suggested that the complex phenomenon which we call 'comfort' has both a *spatial* and a *temporal* dimension. In many buildings the diversity of human activity either demands, or in other instances, may tolerate variation of environment from space to space. It may not be necessary to maintain the whole of the interior of a building at a uniform set of conditions. In the case of the schools which were studied it was evident that energetic activities, such as drama classes or physical education require different conditions to those necessary for academic study in the classroom. Indeed many activities could be satisfactorily accommodated in conditions which were either warmer or cooler than those of a classroom, in other words they were less environmentally critical. In addition to suggesting that buildings may be planned with areas providing different conditions for different activities, there is a further dimension to the idea of spatial diversity. This is that single spaces may be recognised to have simultaneously occurring diversity, and that this may be desirable. For example, some people may seek out a sunny part of a room, whilst, simultaneously, others may prefer to be in the shade. In this way a single space may be able to satisfy the differing needs of individuals or to accommodate different activities.

From these studies it was possible to make a number of observations about the design of control mechanisms. The most important point to be made is that controls which are to be used by a variety of individuals in a building, in other words dispersed user-operated controls as opposed to centralised controls, must be comprehensible. Their purpose must be apparent.

For example heating or lighting controls should clearly be identifiable as such and it should be obvious to the user which equipment they are connected to. This requires particular care in the location of controls so that they will be effectively used when environmental action is required. Whenever people decide that they need to adjust the environment in a building they require a rapid, if not immediate response to their actions, they respond to unacceptable or inappropriate conditions and require instant relief. Switching on artificial lighting is an obvious example of such a response. Other mechanisms, such as some forms of heating system, are less quick to act. This is why the nature of the system must be understood. A further, and most important aspect of user response is that the spur to action to rectify discomfort is more reliable than the memory to return and undo the response when comfort has been restored, and particularly when the need for it has passed. The most familiar example of this neglect is the continuing use of artificial lighting switched-on on a winter's morning and unnecessarily left on after the day has brightened.

2 Theory into practice

The first practical application of these principles was in the design and construction of the Netley Abbey Infants' School in Hampshire [4]. The Hampshire County Architect's Department collaborated with researchers based at the Martin Centre to translate the principles of selective design into guidelines for the design of school buildings. These were then applied in the design of the Netley School (Fig 2).

The essence of the design was the use of passive solar gains collected through the conservatory which runs the full length of the building's south-east face, the distinction between two distinct levels of environmental control, relatively precise in the main teaching spaces and more variable in the conservatory, and the reliance upon the occupants to judge the state of the environment and to take appropriate action in its control. Since its completion in 1985 the building has operated in this mode and, as monitoring carried out during its first two years, 1985-86 and 1986-87, has shown [6], the building has maintained a high quality of environment with great energy efficiency.

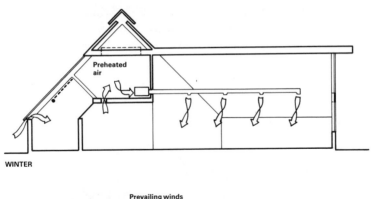

WINTER

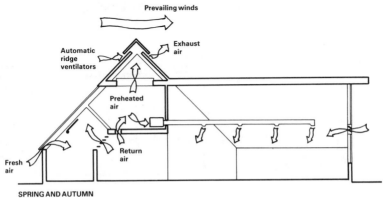

SPRING AND AUTUMN

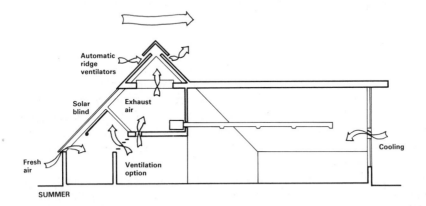

SUMMER

Fig 2   Netley School, Netley, Hampshire

The close collaboration between the research team and the designers of this building, and the derivation of the underlying principles from studies of school buildings almost certainly contributed to the success of the design. But the true test of theory must come when it is applied more widely. Two recently completed buildings allow us to examine the value of "Selective" control in different situations. The first is the School of Engineering at De Montfort University, Leicester, architects Short Ford Associates, [5] and the Friary Project at Maldon in Essex by Greenberg and Hawkes [1].

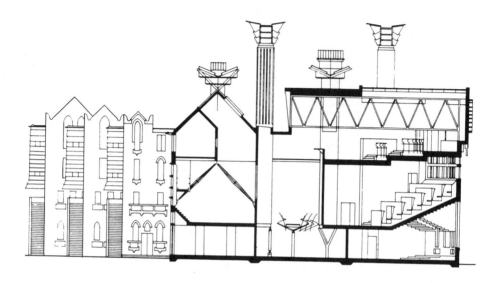

Fig 3 The School of Engineering, De Montfort University, Leicester

The School of Engineering is remarkable for the extent to which it makes use of "passive" processes of environmental control. The most reported element of this is the reliance on natural ventilation of its principal lecture rooms using tall vent shafts which rise through the building to emerge high above its roof line. In the context of the present discussion, however, it is other aspects of the design that merit comment. The first point to note is that the building houses an enormous range of functions. In addition to the large lecture theatres it has smaller seminar rooms and classrooms, teaching laboratories, research laboratories of differing dimensions and purpose, and academic and administrative offices. This means that the users of the building, academic staff, administrative staff, technicians and students, develop a complex set of relationships with the spaces they inhabit. For example, a lecturer will move from his personal office to a seminar room, then to a teaching laboratory, then to a lecture room, and so on. A student may one day attend a series of lectures and another work entirely in a laboratory. Sometimes people will work in large groups, at others in small groups, and yet at others they may work alone. All of these conditions redefine the nature of their relationship to the building and, hence, their role in operating its environmental processes.

On a visit to the building in 1994 the author recorded a number of specific malfunctionings of the environmental system. Two instances concerned the familiar problem of the control of artificial lighting. First, in the mechanical engineering laboratory, a large top-lit space, all of the artificial lights were in use when only a small part of it was in use by a group of technicians. A check of the level of natural light showed that this was insufficient to perform precision work at machine tools, but the use of the entire electric light installation was the result of poor location and design of the switching which made it easier for the users to put it all on rather than to illuminate just the area of the space they were inhabiting. The electrical engineering laboratories are much smaller in scale. Here great care has been invested in the design of their daylighting. Light shelves are used in relatively narrow spaces to provide a uniform distribution of light while protecting workplaces from glare. The artificial lighting is controlled by the occupants and consists of two systems providing general background illumination and local sources at individual workplaces. The general lighting is switched from a single position located in the lobby just outside each laboratory and the local lighting is locally switched. On the visit to the building all of the lights were in use when the laboratory was inhabited by only three students. The daylighting level, when tested by simply switching off the artificial lighting, proved to be more than sufficient as general background.

In both of these examples the question which arises is that of the 'legibility' of the controls. This is a question of their location in or relative to the space and to the devices which they operate. Unless the purpose of controls is clear to the user, particularly to one who is an intermittent occupant of the space, it is unlikely that it will be operated intelligently. Similar problems where recorded in the thermal systems of the building. Even a familiar device such as a thermostatic radiator valve was failing to operate as expected, even in a small private office, where the occupant was responding to overheating on a winter's day by opening a window rather than by setting the thermostat at an appropriate level. This may be because building users have become so accustomed to centralised controls in the workplace that they are reluctant to tamper.

In a large and complex building, such as the School of Engineering, containing such spatial and functional diversity there will inevitably be a corresponding diversity of control systems. In large spaces the primary heating controls will be automated and, as here, be controlled by a centralised Building Management System. In many of the smaller spaces control is vested with the users. This means that they have constantly to change their perception of the way in which the environmental system operates as they move from one space to another. This poses questions of architectural semantics which have, as yet, barely been identified.

The Friary at Maldon consists of two buildings (Fig 4). The first contains a branch of the County Library on the ground floor with two floors of offices for the Social Services Department above. The second, smaller, building is a day-care centre for the mentally handicapped. The buildings incorporate the principles of 'Selective' design in their response to orientation, with larger windows and the major spaces facing south and, in general, smaller windows and secondary spaces to the north. The internal environment is 'free-running' in the summer months, with natural lighting and ventilation. The auxiliary heating is provided by a simple gas-fired system with radiators and thermostatic valves in all parts of the building. A low-energy artificial lighting installation is user-controlled and locally switched. All windows in the library and offices, both north and south-facing, are provided with internal, user-controlled blinds.

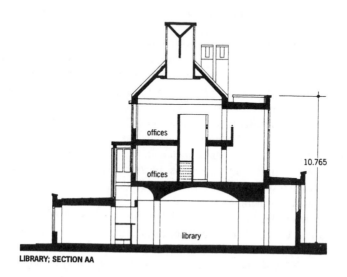

LIBRARY; SECTION AA

Fig 4   The Friary Project, Maldon, Essex

A study of the environmental conditions in the County Library was carried out in the summer of 1994 [7]. As part of this a daylight study model was made of a typical bay of the space. This showed that the daylight factor varies from a maximum of 7.92% beneath the clerestory windows to a minimum of 1.73% at the centre of the space. The average daylight factor is 4.16%, which, using daylight availability curves, suggests that there will be sufficient natural light for the library function for 74% of the occupied hours per annum. In view of the attention which was paid by the architects to daylighting design, and to the satisfactory levels achieved, it is ironic that, at the time of the survey in 1994, under bright summer conditions, the artificial lighting was permanently in use. The senior librarian asserted that this was County Libraries' policy for all its branches in order to ensure satisfactory levels for partially-sighted users.

The control of summer temperatures is achieved by the now familiar 'Selective' or 'passive' devices of thermal mass and natural ventilation. On the day of the user survey the outdoor temperature was 18.8 deg.C at 09.00 and rose to a maximum of 23.9 deg.C at 15.00. The internal temperature was 23.2 deg.C at 09.00 and reached at maximum of 25.2 deg.C at 13.00. The building was quite densely occupied throughout the day, with over thirty staff and readers often present. The modern public library also contains a number of heat-producing devices in the form of computers and photocopiers which contribute to the heat gains. The building is cross-ventilated through high level windows to the north and a clerestory at the south rising above the lower roof of the south wing. Trickle ventilators are provided in the low-level south-facing windows, but opening lights were not permitted because of the risk of loss of books. The ventilation of the building is driven by wind-pressure and a slight internal breeze was noted during the study which was appreciated by the staff.

It was observed that users preferred to sit a the reading places in the most highly daylit parts of the space. Even on a warm summer's day readers were observed sitting in direct sunlight on the wide cills of the south-facing windows.

All of the environmental devices, opening lights, blinds, artificial lighting and thermostatic radiator valves are user operated. In this case this means they are operated by the library staff and not by the readers. The staff, thus, act as monitoring devices on behalf of all users. This is in some way similar to the condition in a school building where the teacher usually performs this role on behalf of the students. In response to interview the staff stated that they found the building easy to understand and operate, although, as reported below, there is one aspect which is unsatisfactory.

In the summer conditions of the study the staff opened all of the high level opening lights when they first arrived in the building in the morning. They explained that they regarded the blinds as part of the building's security system, drawing them in the evening and raising them when they arrived in the morning. During the study, however, the high level blinds to the north-facing windows remained lowered and those on the large south-facing windows were permanently raised, even though there was a considerable quantity of solar gain. The reason for this may be that the blinds are mounted at the front of the deep window reveals where, if drawn, they prevent people sitting on the wide cills.

It is clear that a number of the small-scale details of the building, such as this, compromise the fullest realisation of the architects' environmental intentions in a number of respects. Nonetheless the users interviewed in the study, 10 staff and 21 readers, all expressed satisfaction with the environment. Every respondent found the building 'fresh' and considered that it was 'cool'. The librarians expressed the view that the building is the coolest library in which they had worked .

The study asked the staff and readers to express a view of the winter performance of the building, relying on their memory of just one full heating season in the winter of 1993-94. Their recollection was that the building was 'comfortable' and 'warm' although there had been a few instances of overheating on busy Saturdays. No evidence was available about the ambient conditions on these occasions so that it was not possible to discover if these occurrences coincided with sunny days and, hence, with the presence of solar gains. These enquiries revealed that the staff were unaware that the radiators were fitted with thermostatic valves and so were in a position to control the temperature. There had been no formal induction of the staff on the environmental strategy of the building. In these circumstances it is perhaps remarkable that the users had established such an intelligent and appropriate mode of control over the hot summer environment. On the other hand it is regrettable that the basis of a carefully designed winter strategy had not been explained, although it would appear that this had not led to any environmental problems.

3 The specifics of 'selective control'

The experience of these two buildings, whilst not representing a statistically significant sample, points to a number of issues which require consideration if user control of the environment in buildings of this type is to ensure both acceptable environmental conditions and efficiency in the use of non-ambient energy sources.

Perhaps the first point to be made is that a building is primarily a social system and that its representation as a physical system must take this into account. People go to buildings to work, teach, be taught, borrow a book, and myriad other things, and these purposes are at the forefront of their minds. The control of the environmental systems of the building, whatever their nature must be only a secondary concern. It should, therefore, be easily achieved, without ambiguity, and be effective in producing whatever result is desired. This means that the designers of buildings must understand the users' priorities in great depth if they are to achieve this goal.

In development of the theory of 'Selective' control the following points should be observed by designers:

1. Identify the full range of uses to be accommodated.
2. Make a detailed schedule of activities, accounting for the numbers of users participating in each, the composition of the group, the times when each activity may occur.
3. Design all controls, whether on plant or fabric, to be simple and clearly associated with the building element which they operate, and locate them so that they might be used appropriately.
4. Ensure that the detailed design of elements does not compromise their own or other elements' operation.
5. Inform the users about the controls in the building and how they should be operated.

4 Conclusion

The move to the restoration of user control over the environment in buildings has developed in recent years as part of the desire both to exploit ambient energy sources and to provide better standards of comfort than conventional numerical specifications, with their familiar association with mechanical plant and auto matic control will permit. The performance of such buildings requires constant monitoring if they are to deliver the performance to which they aspire and in order to continue the process of development of the theoretical basis of good design practice. The present paper is a small contribution to this process.

5 References

1. Colbourne, C.   (1994),   Building Study: The Friary Project, Maldon, in *The Architects' Journal*, 15 September 1994.
2. Hawkes, D. and Willey, H. (1977), User response in the environmental control system, in *Transactions of the Martin Centre,*, Vol.2, Woodhead-Faulkner, Cambridge.
3. Hawkes, D. (1982), The theoretical basis of comfort in the "Selective' control of environments, in *Energy and Buildings*, Vol.5.
4. Hawkes, D.   (1988), Building Study: Netley Abbey Infants' School, in *The Architects' Journal*, 22 June 1994.
5. Hawkes, D. (1994), User control in a passive building, in *The Architects' Journal*, 9 March 1994.
6. Martin, C. (1987) *Netley Abey School, Monitoring Report, 1986-87*, Energy Monitoring Company Ltd.
7. Nikolopoulou, M-H, (1994), *Human comfort and occupants' interaction*, M.Phil dissertation, Department of Architecture, University of Cambridge (unpublished).

6 Acknowledgement

The work described in this paper was, in part, undertaken as part of a research project, *The Selective Environment*, at the Martin Centre, Department of Architecture, University of Cambridge under the sponsorship of The Mitsubishi Fund for Europe and Africa.

CHAPTER FIVE

Environmental Criteria for Naturally Ventilated Buildings

Robert R Cohen

Halcrow Gilbert Associates Ltd., Burderop Park,
Swindon, UK

Abstract

Naturally ventilated buildings typically have lower energy running costs and lower capital costs than their mechanically-conditioned counterparts. Many building developers and owners in the UK are keen to construct these environmentally responsible buildings, however they face one key problem: how to demonstrate that satisfactory comfort conditions will be achieved. In the heating season thermal comfort is affected by draughts, asymmetric radiation and vertical temperature gradients. However, it is the risk of overheating in warm weather that is of paramount concern in non-air-conditioned buildings.

A review is given of some of the many standards and guidelines that are used in Europe to specify warm weather comfort criteria, including: the CIBSE Guide, the BRE Environmental Design Manual, Design Note 17, BRECSU Good Practice Guide 71, the International Standard ISO 7730, European Standard CEN 27730, SCANVAC, ASHRAE Standard 55-92, Standard S1A 382/3 (Switzerland) and Overheating Guidelines in The Netherlands.
And the following key questions are addressed:

- What do people experience as thermally comfortable conditions ?
- What measures can be taken to ensure a building is thermally comfortable?
- How can the thermal comfort of a building be assessed at the design stage ?

Keywords: Buildings, environmental criteria, natural ventilation, overheating criteria, thermal comfort.

Naturally Ventilated Buildings: Buildings for the senses, the economy and society. Edited by
D. Clements-Croome. Published in 1997 by E & FN Spon. ISBN 0 419 21520 4

1 Introduction

The current UK Building Regulations relating to Conservation of Fuel and Power (Part L 1995 Edition) concentrate on reducing the energy consumed for the heating of buildings. However, in many buildings, particularly those used for offices, summertime conditions are increasingly more critical as regards total annual energy consumption. This is an inevitable consequence of two factors: building heat losses are being reduced by improving insulation standards and the use of electrical equipment and appliances is increasing. These factors both cause heat demand to be lower in winter and cooling demand or overheating risk to be greater in summer.

It is recognised by the UK government that legislation has by far the greatest impact on energy efficiency[1], ahead of Codes and Standards, consensus guidance and market forces. So, in common with other EC countries, consideration is now being given to including passive cooling measures or constraints on the use of air-conditioning in building regulations. In 1993, BRE proposed amendments to the Building Regulations, which were intended to restrict the application and/or the energy consumption of air-conditioning and mechanical ventilation, but these were rejected in 1994 following representations from the building industry. A new form for the proposed amendments is currently being developed.

Many people within the UK are keen to see buildings constructed which avoid the use of air-conditioning. Naturally ventilated buildings, in particular, typically have lower energy running costs and lower capital costs than their mechanically-conditioned counterparts[2]. They therefore generate lower CO_2 emissions and also completely avoid the need for CFCs. Furthermore, there is a recognition among many building professionals that occupants may actually prefer to work in naturally ventilated environments[3]. This has led one leading commentator to speculate that the prestige and added value associated with air conditioning in the 1980s will be less prevalent in the 1990s[4]. However architects and developers need to be confident at the design stage that conditions within a building will be comfortable without air-conditioning. This means answering the following key questions:

- What do people experience as thermally comfortable conditions ?
- What is unacceptable overheating?
- What measures can be taken to ensure a building is thermally comfortable?
- How can the thermal comfort of a building be assessed at the design stage ?

These questions are tackled in the sections below.

2 What are the thermally comfortable conditions?

Qualitatively, a recent BRE Information Paper[5] states that the best buildings for overall comfort have two key features:

Feature A: that the buildings provide conditions that are within the accepted comfort range for most occupants most of the time (so people do not need to change things that much).

Feature B: that people have the facilities to alleviate discomfort quickly when it occurs

Quantitatively, thermal comfort is a complex issue and difficult to assess with simple criteria. There are six basic variables which affect whether or not an individual feels thermally comfortable at any one time. These are:

Two personal variables

- clothing level
- degree of activity

Four environmental variables

- air temperature
- radiant temperature
- air velocity
- relative humidity

A number of different approaches have been developed to combine the four environmental variables into a single indice for the assessment of thermal comfort. Having decided how to measure thermal conditions, it is then necessary to decide what constitutes acceptable and unacceptable conditions of thermal comfort for a particular building and occupant type.

With air-conditioned buildings it is possible to specify the environmental comfort criteria within defined ranges such as 22°C $\pm$1.5°C and 30%-70% relative humidity. Clearly, buildings without air-conditioning will not be able to provide such closely controlled conditions. However, experience suggests, that, where their function does not entail extreme levels of internal heat gains, buildings can be designed so that summertime conditions are maintained at acceptable levels without resorting to mechanical cooling. But this begs the fundamental question: what is an acceptable level?

Thermal comfort research by the likes of Fanger[6] has identified the average preferred neutral temperature for 'closely controlled' buildings; there is now also a consistency between the data from climate chamber experiments and field measurements in air conditioned buildings[7].

However these target set-points are not applicable to buildings which are "free-running" in summer. In these non-air-conditioned buildings, the conditions tend to vary with the outdoor climate, often to the preference of occupants who are naturally responsive to the latter. Nevertheless, there are limits as to how far people want their working environment to mirror rising outdoor temperatures.

A number of studies, such as that by Wyon[8] have shown that health and productivity as well as comfort suffer to some extent as temperatures climb above, say, the 26°C mark. However, the effects are difficult to quantify, particularly for example in office workers, making it hard to decide when discomfort reaches the point where air conditioning, despite its concomitant extra cost and environmental impact, becomes imperative.

The problem at the design stage is not just to have credible criteria against which to judge conditions in a building but also to be able to predict these conditions. Given the uncertainties in all present prediction methods, it is inevitable that a certain amount of empirical calibration may be necessary to translate what constitutes unacceptable conditions in real buildings into quantitative criteria applicable to design stage predictions.

It would be convenient if unacceptable overheating could be quantified by a series of simple criteria, such as:

- under what conditions of air and radiant temperature, air speed and so on is the thermal environment unacceptable?
- for how long must that threshold be exceeded on a single occasion for it to be unacceptable to the occupants? Is it, for example, five minutes, 30 minutes or one hour?
- how many occasions per year should such unacceptable conditions be reasonably tolerated? 2½% of the working year, or what?

Unfortunately, matters are not that straight-forward, as comfort criteria are affected by other influences. They depend on context: building type, location and the dress culture. Also there is the observed tendency of occupants to trade-off higher temperatures against other benefits in a well-liked building[9], and, as described in the BRE IP 3/95, the willingness of occupants to forgive nominally uncomfortable conditions, when they have individual access to controls or devices which can improve matters, such as openable windows, light switches or blinds[10].

3 What constitutes overheating

3.1 UK standards and guidelines
Various standards and guidelines are in use which use different indices as the criteria for thermal comfort. In the UK these tend to be based on a temperature calculated as a combination of the environmental variables - the equivalent, resultant or effective temperature

The CIBSE Guide[11] states in section A1 that resultant temperature is recommended for use in the UK, and suggests that there will be no significant increase in dissatisfaction so long as the actual temperature is within $\pm1\frac{1}{2}$°C of the chosen value. Table A1-2 in the Guide gives the recommended design condition for summertime occupancy as an internal resultant temperature of 23°C. The Guide describes the effect of air movement on comfort, and implies that air speeds in the range 0.5 to 1.0 m/s would require an increase in the recommended resultant temperature of some 2 to 2.5°C, thus raising the summer design value to 25 or 25.5°C, in these circumstances.

The Guide also suggests that there is a close relationship between the preferred indoor temperature and the mean outdoor temperature for the location and the season of the year.

The Guide is somewhat equivocal on how to define overheating. It indicates in Section A8 that 27°C is the temperature when occupants might start to consider themselves 'uncomfortably hot'. Section A8 goes on to describe a manual method to calculate at the design stage the peak environmental temperature, and recommends that weather data giving a 2.5% level of design risk are employed. Unacceptable overheating would thus appear to be defined as an environmental temperature over 27°C for more than 2½% of the year.

The BRE Environmental Design Manual[12] is a design guide for naturally ventilated buildings. It suggests the following standards for summertime temperatures:

Standard	Temperature over the working day
'satisfactory'	23°C ± 2°C
'intermediate'*	24°C ± 2°C
'minimum acceptable'	25°C ± 4°C

* the standard agreed for offices of the Government civil estate.

Design Note 17[13] provides guidelines for environmental design and fuel conservation in educational buildings. It states that during the summer the recommended design resultant temperature should be 23°C with a swing of not more than 4°C about the optimum. It goes on to say: *"It is undesirable that the resultant temperature should exceed 27°C during normal working days over the school year but an excess for 10 days during the summer is considered a reasonable predictive risk"*.

BRECSU Good Practice Guide 71[14] provides advice to building clients and their consultants on choosing an air-conditioning system. However it first states that comfort cooling is not necessary if one can answer yes to the following question: *"Will it be acceptable for your space to reach 28°C on the few hours each year when outdoor temperatures reach this level?"*

It goes on to state: *"Even in the warmest parts of the UK, outdoor temperatures exceed 28°C for only 10 hours and 25°C for only 40 hours in a typical year. Even when the outside temperature is 28°C, the temperature inside the building may well be lower"*.

"Discomfort inside buildings in hot weather is compounded by factors other than just temperature and high humidity. Lack of air movement and direct sunlight can also contribute to discomfort. A well designed ventilation system which can include opening windows and external solar screening can overcome this. Thus, in many situations, the need for air conditioning can be eliminated altogether by removing the other causes of discomfort".

A BRE Discussion Paper (Report No XXX[15]) states: *"It would be reasonable to provide cooling for comfort purposes if a design study shows that heat gains to rooms would give rise to a dry resultant temperature likely to exceed 28°C for more than 1 hour on 10 or more weekdays per year, and that these gains cannot be removed satisfactorily by natural or mechanical ventilation"*.

3.2 International publications

Many of the international standards use an approach developed by Fanger based on experiments conducted in climate chambers[16]. They employ the concepts of predicted mean vote (PMV) and predicted percentage dissatisfied (PPD). The PMV is the predicted mean vote, on the ASHRAE seven-point thermal sensation scale, of a large group of persons when they experience a particular combination of air temperature, mean radiant temperature, humidity and air velocity with their metabolic rate and clothing insulation at given levels. The mathematical relationship between PMV and the six determining variables was derived statistically from the experiments, but is based on a heat balance equation for the human body solved for a range where sweat rate and mean skin temperatures are within comfort limits. An individual is deemed to find conditions acceptable if his vote is one of the three middle points of the scale. However variance between individuals, and interestingly between the same individual on different occasions, means that even when the PMV is at the centre of the scale, statistically 5% of the group will have voted outside the central three points and are therefore considered to be dissatisfied.

The International Standard ISO 7730[17] and the European Standard CEN 27730[18] are very similar thermal comfort standards based on PMV and PPD. Given the observed impossibility of providing conditions which will satisfy all people all of the time, the Standards adopt the criterion that 10% uncomfortable (PPD = 10%) is acceptable as a working maximum. This means that the PMV must be within 0.5 of the central point of the scale which translates to an acceptable tolerance on the optimum temperature of ± 1.5K.

The recommended upper limit for operative temperature during summer conditions is 26°C, assuming light mainly sedentary activity and a mean air speed of less than 0.25 m/s.

ASHRAE Standard 55-92[19] is in close agreement with ISO Standard 7730. It recommends limits for general thermal discomfort should be based on a 10% dissatisfaction criterion which leads to an upper limit for operative temperature in summer of 26°C, assuming 50% relative humidity, sedentary activity, 0.5 clo and a mean air speed of less than 0.15 m/s.

SCANVAC[20] constitutes a voluntary code of practice for application in Scandinavia. It provides a classification of the standard of an indoor climate with respect to thermal comfort, air quality and noise level. Thermal comfort performance is based on PMV and PPD. Summer overheating is defined in terms of operative temperature with upper limits of 25.5°C, 26°C and 27°C respectively for the three defined standards (in order of quality), with the first assuming individual control over temperature and airflow. These values assume light clothing (0.5 clo) can be worn.

Standard SIA 382/3[21] applies in the Zurich Canton of Switzerland. It requires at least one of the following circumstances to exist in order for the installation of air conditioning to be permitted:

- Special requirements eg clean rooms, computer centres etc.
- High internal loads, from lighting, equipment and people. Levels of internal loads which justify air-conditioning are quoted for various situations.

- If it cannot be proved that air conditioning is required using the above criteria, it can still be proven by showing that the indoor temperature exceeds 28°C for at least 30 kelvin hours/year. Note that periods when the ambient temperature exceeds 30°C are excluded.

The Dutch have had nominal criteria for maximum summer temperatures for many years. Guidelines published by Rijksgebouwendienst (RGD) in 1979[22] stated that "the maximum allowable indoor air temperature is 25°C, in the case of relatively high ambient temperatures. During about 5% of the annual working time the indoor air temperature is allowed to exceed this 25°C level. During 1-2% of the annual working time the indoor air temperature is allowed to exceed the 28°C level, but only occasionally and only caused by extreme ambient conditions".

In 1991, the RGD published a new method[23], the 'weeguren methode', which calculates the duration of any overheating (defined as a PMV of 0.5 or higher) and multiplies it by a weighting factor dependent on the PMV. For PMV = 0.5 the factor is 1, for PMV = 1 the factor is 2.5. The recommended maximum number of weighted hours is 150 weeguren, which is equivalent to, for example, 100 hours at a PMV of 0.7 (which has a weighting factor of 1.5).

3.3 Thermal balance and adaptive models

The climate chamber based, thermal balance model of thermal comfort was originally developed to enable the HVAC industry to provide satisfactory indoor climates in open-plan (ie occupied by a group) air-conditioned spaces. The latest consensus is that this application of the approach is both valid and useful[24].

A radically different model of thermal comfort, known as the adaptive model[25], is applicable predominantly to buildings which are "free running" in summer, that is the indoor conditions are allowed to vary with the outdoor climate.

Fanger in a recent paper[26], indicates that he remains deeply sceptical of the principles of the adaptive model, particularly in its application to office buildings in developed countries:

"Nicol and Humphreys have gathered data from a substantial number of field studies in various parts of the world. For "free-running" buildings, ie buildings not heated or cooled, they found a remarkable relation between preferred temperature and average monthly outdoor temperature. People seem to be reasonably thermally neutral over a large interval of indoor temperatures. Obviously people have altered their clothing and maybe also their activity to maintain reasonable thermal neutrality even at quite high or low temperatures.

In addition there may also occur some physiological adaptation which would require that the occupants experience cool or warm discomfort, probably for weeks. Nicol has the intention to develop an "adaptive model" for thermal comfort. The idea is that people gradually should adapt to the temperatures that happen to occur in free-running buildings. The application of this idea provides an important energy conservation potential. This idea may very well work in dwellings for people who desire to save and also are ready to suffer a certain discomfort during the adaptation process.

It will be interesting to follow the progress of Nicol's studies but the idea of adaptation is in contradiction to the basic rule in ergonomics: that the machine should be adapted to the human. In contrast to this it is Nicol's idea that the human should adapt to the machine (the building). This principle, especially the physiological adaptation is probably less likely to be acceptable in office buildings".

Although there is little dispute that the thermal balance model is currently the best available objective predictor of comfort for a fixed set of room conditions, it is doubtful whether its inherently steady-state foundations can allow it to be applied reliably to the transient situation in free running buildings. Even so, Fanger does not appear to rule out the application of his model to the occupants of free-running buildings and has put forward explanations for apparent discrepancies[27]. Nevertheless, a more promising approach would appear to be the "adaptive opportunities" model proposed by Baker and Standeven[28]. In part this combines the thermal balance and adaptive models. It assumes that as the environmental conditions vary, the occupants will, if possible, attempt to maintain a thermal balance by adapting to such changes, as a minimum through their posture, clothing and activity level, and possibly by adjustment of, for example, shading devices, openable windows etc. In principle, this means that a well-designed non-air-conditioned building in the UK with a high degree of adaptive opportunity may be able to maintain the PMV within 0.5 of the central point of the scale at most times. This may incidentally indicate an apparent rise in preferred indoor *temperature* as the outdoor temperature increases. However the underlying reasons for such a correlation may rest more with the occupants' ability to attain thermal balance by adjusting other comfort parameters than with any innate acclimatisation to outdoor climate.

An important pragmatic question, particularly in relation to verifying a performance guarantee, is whether comfort can be measured objectively in the varying conditions of a free running building. This needs further work. Equally, the prediction of PMV in free running buildings at the design stage is not easy (see section 5). If these problems can be overcome, the principles of the new Dutch guidelines, the weeguren method, appear to be the best way forward. They embody both the thermal balance theory and a key element of the adaptive model by recognising the inevitable fluctuations in the summertime climate, and allowing for a given minimum level of overheating during short periods of hot weather.

3.4 Evidence from real buildings

The thermal comfort of occupants in any building is best assessed subjectively, however, despite the huge resource that has been applied to thermal comfort research since the pioneering survey of the comfort of factory workers in England by Thomas Bedford in 1936[29], there is a dearth of evidence from the occupants of real buildings concerning what they would consider constitutes unacceptable overheating in summer.

This is particularly the case if one is considering specific building categories such as, for example, non-air-conditioned, office buildings, in the UK, where the survey has been done during the summer and includes some hot weather spells. A further shortcoming of most of the field measurements is that only temperature is reported, and not the five other thermal comfort variables.

A thorough but by no means exhaustive search of the literature has yielded the following studies which are considered directly applicable to UK non-air-conditioned office buildings.

Fishman and Pimbert[30] conducted the Watson House survey in the late 1970s. This found that the average neutral globe temperature in summer was 21.9°C, when the average actual globe temperature was 23.3°C. Furthermore the study indicated that people are more critical of warmer-than-neutral conditions than Fanger's comfort equation would suggest.

Measurements by Croome et al[31] in an office at the University of Reading in early summer (thus missing the main summer period) found a close correlation between the measured neutral temperature and that predicted by Fanger's equation.

Griffiths[32] took measurement in 9 office buildings (8 in the UK, 1 in Germany), but only 4 of these surveys were done outside the heating season, and only one of those (for the German offices) was in mid-summer. Griffiths found that the preferred operative temperature was some 2-3°C lower than that predicted by the Fanger model, but no discussion of this finding is provided in the report.

The Energy Performance Assessment (EPA) project involved the detailed monitoring of 30 UK "passive solar" buildings of which 3 were non-air-conditioned office buildings; case studies have been published for 2 of these: South Staffordshire Water Company[33] and Gateway II[34]. In the former, maximum recorded internal temperatures in the year March 1987 to February 1988 inclusive were about 24°C and an occupant survey found that precautions against summertime overheating worked well. In Gateway II, office temperatures were reported to be always below external temperatures in the 12 months from April 1990 to March 1991 inclusive. This period included very much hotter than normal summer months and "occasional problems of overheating" are noted in the amenity assessment section of the Summary Report. The far more detailed Final Report contains the following assessment of summertime performance during this exceptional year:

Most people were generally quite satisfied with the thermal comfort in winter, with only 15% expressing dissatisfaction. Responses to the thermal comfort in summer, however, were equally spread between positive and negative, the median rating was therefore slightly worse than for winter, with 34% expressing dissatisfaction.

About 80% of the respondents said that summertime overheating was a problem to some degree. The median response however, indicated that it was only a problem "a little".

Measurements during what was a very hot summer showed that the office temperatures were often around 25°C and rose to 27/28°C on the south and west sides during the afternoon. Note though, that during normal working hours, the internal temperature never rose above the external.

During the warmest period of the year the external temperature rose to about 35°C during the working day, the maximum office temperature, recorded on the south side, rose as high as 31°C, with the average office temperature about 2°C lower. Throughout the working day, the office temperatures were typically between 3 to 5°C below the external.

Analysis was performed to determine if there was any relationship between reported summertime overheating and any of the following: floor level, position within the office (ie close to an external window, close to the atrium, mid position in the office, or in an external corner with two sides of windows), or whether the respondent worked in an open plan or private office; however none was found.

A new series of field studies looking at the summertime conditions in occupied naturally ventilated and mixed-mode buildings has recently been reported[35]. One conclusion was that *"suitable and properly-controlled combinations of thermal mass, mechanical ventilation and openable windows were able to keep peak internal temperatures below those outside"*.

However, it also contained the following warning: *"In naturally-ventilated buildings, occupants are often expected to accept higher peak temperatures (typically up to 27°C) than in sealed ones. The case studies supported this, but only in good buildings with good outside awareness. Where there was poor outside visibility (eg views and air flow blocked by partitions and screens), deeper plans (ie occupants remote from windows and not in control of them, or adjacent to atrium windows), gloomy lighting (eg LG3 Category 1 lights plus dark furniture and screens), or poor control of ventilation (eg poor window adjustment facilities or desks in the way), complaints of overheating could occur at 25°C or less. Designers must not assume that just because they happen to have provided natural ventilation, occupants' tolerance levels will necessarily increase."*

Another potentially rich source of experience might have been The BRECSU Best Practice Programme case studies, however these focused primarily on energy consumption, and generally did not involve any detailed monitoring of summertime temperatures even in the non-air-conditioned buildings. Nevertheless the case study reports do contain some qualitative assessment of the occupants views of their summertime comfort conditions, and a summary of this admittedly anecdotal evidence is given below.

Good practice case study 1, Policy Studies Institute, London
"The design solution allowed many of the rooms to be naturally-ventilated, with mechanical ventilation to the atrium and surrounding offices only, and to conference and meeting rooms on the ground floor."

"In summer the conference suite gets rather warm, and this is exacerbated by the radiation from the tungsten-halogen lighting. PSI have considered adding cooling to the conference room ventilation system, but as at the outset the price was too high."
"The peak summer temperatures in the conference suite have been disappointing. Without resorting to air-conditioning, potential remedies could include:

- *using more efficient light sources*
- *omitting the suspended ceiling*
- *programming the ventilation system to run overnight"*

Good practice case study 13, NFU HQ, Stratford-upon-Avon
"Summertime temperatures are acceptable in spite of more electronic office equipment than was foreseen at the time of design."

"Offices mechanically ventilated to 2 ach (4 if necessary in summer), with openable windows. Top floor air conditioned with minimum fresh air and fan-coils. Separate comfort cooling systems for management dining, board room, conference room, training rooms and computer suite."

"The use of thermal capacity has been very effective, with the building exhibiting a higher degree of thermal stability than had been calculated, the hottest parts peaking at only 27°C in the abnormally hot summer of 1989."

"The design team initially anticipated that the offices would require some refrigeration summer, but computer modelling indicated that this would not be necessary with good thermal capacity, solar gain control, daylighting, and low capacity mechanical ventilation. However at that time the allowances made for office equipment heat gains were smaller than today, when more contingency provision for refrigeration would probably be made."

Good practice case study 14, Cornbrook House, Manchester
"Generally naturally ventilated, with tempered mechanical ventilation to toilets."

"User Reactions: the building is generally regarded as satisfactory. A few occupants commented on insufficient ventilation, possibly a reaction to the small windows, low natural light level, and fairly warm environment. The designers would like to have included background mechanical ventilation but it was too expensive."

Good practice case study 15, Hempstead House, Hemel Hempstead
"The naturally-ventilated offices are 16 metres wide and largely open-plan. Packaged split-system air conditioning units are fitted in the conference room, print room, and tenant's machine room only. These are operated on-demand under their own local temperature controls."

"There have been a few complaints of summer overheating due to the generous areas of glazing on the west elevation."

Good practice case study 19, South Staffordshire Water Company, Walsall
"Overheating has not been a problem, even in the hot summer of 1989." This corroborates the EPA finding reported above.

"Most people found summertime temperatures satisfactory."

"The new building meets SSWC's objectives and confirms that air-conditioning was not necessary for their requirements."

Good practice case study 20, Refuge House, Wilmslow, Cheshire
"The good thermal performance of the fabric has limited summertime temperatures and cooling requirements. Because the air conditioning does not run everywhere all the time, energy costs are much lower than for most fully air conditioned buildings."

"The combination of natural and mechanical systems has not been without its problems, particularly in hot weather when people are uncertain whether to open the windows or to rely upon the air conditioning. In hot weather the natural impulse is to open windows, instead of keeping them shut and relying upon the cooling system."

General information leaflet 10, Magnus House, Bridgwater
"Ventilation is natural through windows expect for the toilets, which have mechanical extraction. No air-conditioning is present or necessary, in spite of high densities of electronic equipment in several rooms. The ingress of noise and dirt through open windows was not seen as a problem.

"Summertime temperatures have been acceptable, and much more comfortable than in Magnus' former building, showing the benefits of thermal capacity in reducing peak temperatures."

Generally information leaflet 12, Posford House, Peterborough.
"The need for air conditioning was avoided by a combination of thermal mass and a well-designed, fresh air mechanical ventilation system, which uses passive cooling in summer."
"In summer mode the heating is turned off and fans are run at high speed to provide six air changes per hour."
"Staff are very satisfied with the fresh working environment and the summertime conditions which are more comfortable than those in their adjacent 1970s office."
"The thermal capacity of the building helps to stabilise temperatures and reduce overheating in summer."

4 How can designers ensure a naturally ventilated building is thermally comfortable in summer ?

The design of a building strongly affects summertime thermal comfort and overheating risk. Acceptance of the need for air-conditioning can lead to the adoption by the design team of the totally climate rejecting style of architecture epitomised by deep plan buildings with highly glazed facades of solar control glass. Alternatively, if air-conditioning can be demonstrated to be unnecessary the design team can pursue the low-energy climate-responsive approach which embodies measures such as:

*   insulation of the envelope, to prevent conduction gains resulting from the incidence of solar radiation on its outer surface; light colours of the outdoor surfaces are also desirable, particularly on the roof;
*   effective and adjustable shading of transparent surfaces to prevent the penetration of direct solar radiation; south and north-facing glazings are preferable, with east, west, south-east, south-west and horizontal orientations to be avoided as much as possible, for two reasons: the amount of incident radiation and greater difficulty in providing efficient shading;
*   well daylit spaces, daylight and occupancy responsive lighting controls and low installed load lighting systems (e.g. 2.5W/m² per 100 lux);
*   use of high levels of exposed thermal mass, purged by nighttime ventilation, to moderate indoor temperature diurnal swings;
*   provision of secure, operable openings to allow for natural ventilation when the outdoor ambient temperature is lower than the indoor temperature (including during the night under BEMS control);
*   tall floor to ceiling heights to provide a reservoir for warm air above head height and to enable deeper penetration of daylight from high head windows.

The results of monitoring a range of 'lower-energy offices' have recently been published[36], identifying many practical design lessons for architects and engineers alike on how best to implement such measures and avoid pitfalls.

In many circumstances, it is possible that the most energy efficient building services can include a mixed-mode approach to ventilation. This might mean that in winter windows are closed (avoiding excessive infiltration and cold draughts) and the minimum fresh air requirements are met by mechanical ventilation often with heat recovery. In summer natural ventilation via openable windows etc. is available during the day (avoiding excessive fan power for free cooling) whilst during summer nights the mechanical ventilation system can provide a secure and controllable means for night purging.

5 Assessing the thermal comfort of a building at the design stage

There is one key problem in ensuring that buildings without air-conditioning are constructed where possible, that is to demonstrate to clients, financiers and their agents at the design stage that satisfactory comfort conditions will be achieved when the building is occupied. All too often air-conditioning is specified just in case the building might overheat in summer. What is needed is an authoritative practical assessment procedure with a credible set of overheating criteria which can be applied to non-air-conditioned buildings at the design stage in order to check whether acceptable comfort conditions will be achieved.

In the UK CIBSE provides a calculation method but only limited guidance as to what limits should be set for acceptable thermal conditions. Calculations can be done by hand or using computer simulation models. Hand calculations are feasible but limited, slow and cumbersome. Increasingly, thermal simulation models are set up in order to examine the performance of a building design.   Models can allow the temperature in every zone of the building to be predicted for every hour of the year assuming various operating conditions. Prediction of PMV in free running buildings at the design stage is not easy, but computer aided techniques are developing fast, specifically the integration of dynamic thermal models with daylight and air flow models, so that direct calculation of PMV as a variable in both space and time should soon be feasible.

The accuracy and reliability of thermal models has been a source of considerable debate and research[37] Comprehensive analytical validation tests and inter-model comparison exercises over the last decade have enabled program developers to weed out coding errors and to ensure most of the algorithms of physical processes contained in each model are robust and reliable. However the results of meticulous empirical validation[38] recently carried out by EMC/De Montfort University/BRE as part of IEA Annex 21, comparing the predictions of models with the measured behaviour of outdoor test cells confirm that thermal models at their present state of development cannot produce exact predictions of how real buildings will behave in practice.

It must therefore be recognised that no calculation method can predict exactly what will happen in a building under actual operating conditions. All methods involve assumptions and none can give 100% accurate results. Even experienced users can produce widely varying results for the same building if the methods they use to model the building and the assumptions made vary, although this variation can be reduced by the use of a Performance Assessment Method (PAM)[39]. In this regard, it is salutary to read the closing paragraph of a recent paper by Wijsman[40], reflecting on his experience with implementing the Dutch overheating assessment method:

"a check of the input files by a second person and the use of a PAM lead to very close results between users that use the same program (using different programs will lead to a bigger scatter). However, these very close results have still a scatter of 30-40 h. In the Netherlands the design of a building is judged on the number of overheating hours: limits are 100 h above 25 C and 20 h above 28 C. In practice this means 99 h above 25 C is adequate and 101 h not!! Having seen the remaining scatter in the results a review of 'regulations' seems to be necessary."

The predicted level of overheating risk will depend on assumptions made for variables such as the level of equipment heat gain, occupant density, etc. Ventilation rates and solar shading effectiveness may also be uncertain. Sensitivity analysis can help to reveal the robustness of the building, but the amount of effort required to generate comprehensive results is considerable.

Overheating risk will also be significantly affected by the weather data employed. Undoubtedly many people would recognise the pragmatism of needing to select one particular weather year for an overheating assessment. The question is whether that year should be average (however that is defined) or extreme. Do clients want to know the level of temperatures in an average (say median) year or in the hottest in 40 years for a 2½% level of design risk? Alternatively, methods may be developed to extrapolate the results from 1 year to say 20 year performance using long term weather statistics. Answers to these questions require detailed examination, a task currently being undertaken by the CIBSE Weather Task Group.

Any assessment of overheating risk needs to be understandable by clients, which means presenting the results in a meaningful format. Options include frequency distributions for zone temperature, PMV, etc. Additionally, it is important that the client understands the level of design risk involved: how often is the building, or parts of it, going to be at a nominally unacceptable condition (eg temperature) - but equally understanding that reducing the level of design risk means, generally, increases in capital cost, increases in energy running costs and increases in management and operating costs. Clearly a balance is required which may vary according to building type, location and the business of the occupants.

Despite the difficulties and uncertainties a standardised calculation method and overheating criterion applicable to buildings in the UK would be useful in order to compare design options and indicate the level of performance that can be expected from a building. If builders and developers can be convinced that buildings which meet the criteria will be thermally comfortable in all but exceptional circumstances, it may stop the tendency to specify air-conditioning as standard without considering alternative approaches.

6 Conclusions

- There is a growing body of evidence that, where location permits, a well-designed building with a high degree of adaptive opportunity, including natural ventilation via openable windows in summer, can be preferred by occupants to a sealed air-conditioned alternative.
- Occupants do not like the thermally uncomfortable conditions which are likely to occur in naturally ventilated buildings during unusually hot weather, but they are prepared to forgive such conditions if they only last for short periods, because of their preference in normal weather for natural ventilation (in non-aggressive locations).
- The thermal comfort of occupants in any building is best assessed subjectively, but, if an objective measurement is required, PMV, based on Fanger's thermal balance model, is currently the best available predictor.
- The overheating assessment used in the Netherlands, the weeguren method, employs criteria which are based on PMV and which allow for a certain amount of overheating which might occur during short periods of hot weather. These advantages make it a prime candidate for consideration as a standard assessment procedure which would provide a consistent, authoritative and practical means to determine at the design stage whether satisfactory comfort conditions will be achieved when a building is occupied. Such a procedure is urgently required in the UK to help clients decide not to specify air-conditioning just in case the building might overheat in summer.

7 References

1. Shaw, M.R. and Treadaway, K.W. "The energy-related environmental issues (EnREI) research Programme", Proceedings of BEP'94 Conference, York, April 1994.
2. BRECSU, Energy Consumption Guide 19, Best Practice Programme, BRE,   October 1991.
3. Duffy F. "Designing comfortable working environments based on user and client priorities", Thermal Comfort: Past, Present and Future Conference, BRE, June 1993.
4. Bordass, W, Comment on "Using natural light and air flow", The Architects' Journal, 28 April 1993, p58.
    BRE Information Paper IP 3/95, February 1995.
5. Fanger, O. "How to apply thermal comfort models in practice", Thermal Comfort: Past, Present and Future Conference, BRE, June 1993
6. de Dear, R, "Outdoor climatic influences on indoor thermal comfort requirements", Thermal Comfort: Past, Present and Future Conference, BRE, June 1993
7. Wyon, D, "Assessment of human thermal requirements in the thermal comfort region", Thermal Comfort: Past, Present and Future Conference, BRE, June 1993
8. Energy Performance Assessment (EPA) Project for the Energy Technology Support Unit (ETSU), 1992
9. Bordass W and Leaman A. "Control strategies for building services", Advanced Climatisation Systems Seminar, Barcelona, June 1993
10. CIBSE Guide, Volume A "Design Data", CIBSE, London, 1986
11. Petheridge, P. Milbank, N.O. and Harrington-Lynn, J. 1988, "Environmental Design Manual: summer conditions in naturally ventilated offices", Building Research Establishment, Watford, UK
12. Department of Education and Science, 1981, Design Note 17, "Guidelines for environmental design and fuel conservation in educational buildings"
13. BRECSU Best Practice Programme Good Practice Guide 71 "Selecting air conditioning systems", BRE, 1993
14. BRE discussion paper: "The use of air conditioning and mechanical ventilation in non-domestic buildings ", Draft for consultation, BRE, 30 November 1993
15. Fanger, P.O. "Thermal comfort: analysis and applications in environmental engineering", Danish Technical Press, Copenhagen, 1970
16. ISO. International Standard 7730, "Moderate Thermal Environments: Determination of the PMV and PPD Indices and Specification of the Conditions for Thermal Comfort", Geneva, International Organisation for Standardisation, 1993
17. CE. EN 27730, "Moderate Thermal Environments: Determination of the PMV and PPD Indices and Specification of the Conditions for Thermal Comfort". Brussels, Central Secretariat - Comite Europeen de Normalisation, 1994
18. ASHRAE. ASHRAE Standard 55-92, "Thermal Environmental Conditions for Human Occupancy". Atlanta, American Society of Heating, Refrigerating, and Air-conditioning Engineers, 1992
19. SCANVAC. "Classified Indoor Climate Systems". Stockholm, The Federation of the Scandinavian HVAC-Organisations, 1991

20. Zweifel G: "Overheating risk assessment - the Swiss experience", Proceedings of BEP'94 Building Environmental Performance: Facing the Future, University of York, April 1994

21. Rijks Geneeskundige Dienst, 1979, "Aanbevelingen voor de arbeidsomstandigheden in kantoren en gelijksoortige ruimten voor de huisvesting van Burgerlijk rijksoverheidspersoneel"

22. Methode voor de deoordeling van het thermische binnenklimaat, Rijksgebouwendienst, June 1991

23. de Dear, R.J. "Outdoor climatic influences on indoor thermal comfort requirements", Thermal Comfort: Past, Present and Future - Proceedings of a BRE Conference, June 1993.

24. Humphreys M.A. "Field Studies and climate chamber experiments in thermal comfort research", Thermal Comfort: Past, Present and Future - Proceedings of a BRE Conference, June 1993

25. Fanger, P.O. "How to apply models predicting thermal sensation and discomfort in practice", Thermal Comfort: Past, Present and Future - Proceedings of a BRE Conference, June 1993

26. Fanger, P.O. "Don't get too comfortable", letter to Building Services, CIBSE Journal, November 1992

27. Baker, N. and Standeven, M. "Pascool thermal comfort studies", Proceedings of European Conference on Energy Performance and indoor climate in buildings, Lyon, France, November 1994

28. Bedford, T. " The warmth factor in comfort at work: a physiological study of heating and ventilation". Industrial Health Research Board Report No 76, HMSO, London, 1936

29. Fishman, D.S. and Pimbert, S.L. "Survey of subjective responses to the thermal environment in offices". In proceedings of Indoor Climate, WHO Conference, Copenhagen, 1978

30. Croome, D.J., Gan, G. and Awbi, H.B. "Field evaluation of the indoor environment of naturally ventilated offices", 13th AIVC Conference, Nice, France, September 1992

31. Griffiths, I.D. "Thermal comfort in buildings with passive solar features: field studies", Report to the CEC, EN3S-090-UK, 1990

32. Solar Building Study 4, Summary Report, South Staffordshire Water Company, ETSU S 1160/SBS/4, 1991

33. Solar Building Study 11, Summary and Final Reports, Gateway II, ETSU S/1160/SBS/11

34. Bordass, W.T. et al, "Naturally ventilated and mixed-mode office buildings: opportunities and pitfalls", Proceedings of CIBSE National Conference, Brighton, October 1994
    Bordass, B. "Avoiding office air-conditioning", Architects' Journal, pp37- 39, 20 July 1995

35. Lomas, K. J. "Thermal Program Validation: the Current Status", BEP '94 Conference, York, 1994

36. Martin, C. & Dolley, P. "The Role of Test Room Studies in the UK Passive Solar Programme", BEP'94 Conference, York, 1994

37. Wijsman, A.J. "Building Thermal Performance Programs: Influence of the Use of a PAM", BEP'94 Conference, York, 1994
38. Wijsman, A.J. " Building thermal performance programs: influence of the use of a PAM", BEP'94 Conference, York, 1994

CHAPTER SIX

Against the Draft CEN Standard: pr ENV 1752

Michael Wooliscroft
Building Research Establishment, Watford, Herts, UK

Introduction

The UK opposition to the document covers four main topics; technical criticism of the approach of the document; the question of indoor air quality, in particular the olf and decipol; the relationship of research and standards; and the possible contractual implications of the standard were it to be implemented even as an ENV - a trial standard. It has also been pointed out that the document does not comply with the PNE rules and significant editing would be required to identify requirements.

Technical Criticism

The technical criticism arises from the requirement that quantities to be measured in the completed system and buildings must be predictable in the design process. Two areas in the proposed ENV where this was not the case were turbulence intensity and indoor air quality, particularly the olf and decipol. A consequence of this requirement is that there must be readily available "catalogue" data on both components and materials. Whilst both of these criteria are in the informative annexe, the distinction between the main document and the informative annexe, whilst clear to those involved in standards, is likely to become blurred to practising engineers. It has been noted that the document is frequently anomalous in that variables which can be controlled are specified with tolerances whereas variables which cannot be controlled are specified without tolerances.

Naturally Ventilated Buildings: Buildings for the senses, the economy and society. Edited by D. Clements-Croome. Published in 1997 by E & FN Spon. ISBN 0 419 21520 4

Indoor Air Quality

Whilst the olf and decipol methodology is contained in the informative annexe it does occupy a significant proportion of the document and if the document were to be adopted as an ENV it would be likely to be used. The olf and decipol methodology has been subject to many criticisms, e.g. Oseland and Raw reference 1, but probably the most cogent criticisms are those emerging from the European Audit Project. The European audit project, more fully the European Audit Project to Optimise Indoor Air Quality and Energy Consumption in Office Buildings, is a survey of some 58 office buildings in 10 countries, involving a range of measurements, physical measurements, ventilation measurements, measurements of pollutants, questionnaire and perceived air quality measurements using the olf and decipol methodology. As can be seen from figure 1 there is no correlation between perceived air quality and Sick Building Syndrome symptoms, from figure 2 there is no correlation between perceived air quality and Total Volatile Organic Compounds (TVOCs) and figure 3, only a poor correlation between perceived air quality and airflow rate.

Figure 1 - Mean SBS Symptoms versus decipol

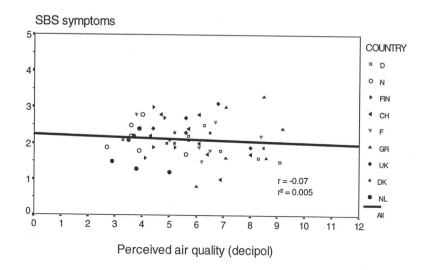

Figure 2 - Decipol versus TVOC

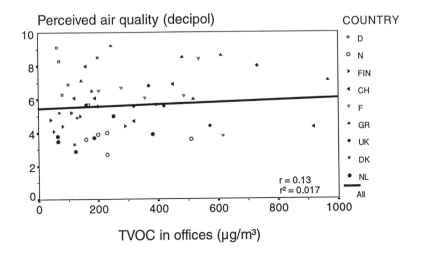

Figure 3 - Decipol versus air flow rate (in all offices)

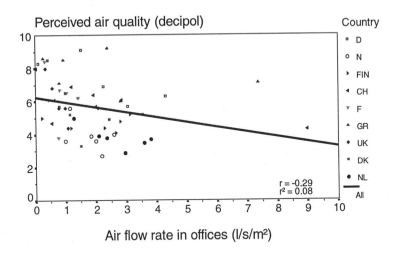

To be fair however, the latter improves considerably when only the UK, Denmark and the Netherlands are considered, where elaborate training procedures of the sniffing panels were undertaken, figure 4. However, the prediction of ventilation

rate is still poor, figure 5 and furthermore a curve fitted to the results gives an
even better correlation, suggesting that beyond a certain point increasing air flow
rate actually reduced perceived air quality.

Figure 4 - Decipol versus air flow rate (in NL, DK and UK offices)

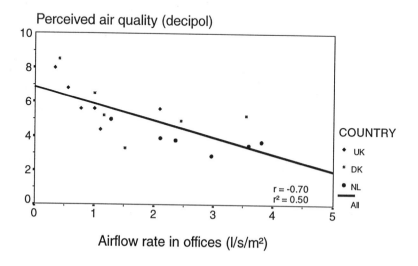

Figure 5 - Confidence limits for decipol versus air flow rate

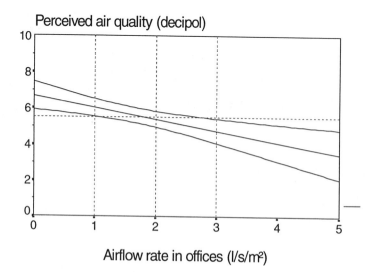

Clearly something strange is happening which would not be expected from the theory of the olf and decipol methodology. The work of Knudsen, Reference 2, also suggests that olfs of different materials cannot be added simply and that two numbers and a logarithmic relationship are involved. This can be dealt with in principle but it is clearly a move away from the simple methodology given in the draft ENV and would require modification to the document.

Research and Standards

Finland has stated on many occasions that "a standard should reflect the state of the art, not the future". The relationship of research to standards is a difficult issue and probably differs between large countries where inevitably the process is more remote and bureaucratic and small countries where researchers and practitioners are much more members of the same community The European Union is effectively a large country writ even larger and it is difficult given also the differences of practice between different counties, differences of meaning of words and terminology to incorporate anything into a standard which is not accepted practice over the major part of Europe. It is suggested that the timescale for the translation of research into practice is probably something like that shown below:

RESEARCH →	PRACTICE →	STANDARDS & REGULATIONS
5-10 years	5-10+ years	~ 5 years

Overall one is probably talking about a generation for research to get into standards.

The Contractual Aspect

In order to design and construct systems to meet specified requirements engineers need comprehensive and reliable data about materials and components. In the absence of such data there is a real risk that consulting engineers and contractors will be held legally responsible for matters over which they have no control. Whilst much of the contentious material is in an informative annexe, there is nothing to stop a client from incorporating this material into a specification. Furthermore, table 1 of the main document says that it is based on the assumption of low polluting materials, but nowhere in the standard are low polluting materials defined, nor any guidance given as to how the designer may find out what are low polluting materials. It was pointed out in the discussion that in the case of industrial ventilation of known pollutants, there is data available on emissions and production rates of these pollutants. Little or no such data is available for building materials.

Where do we go from here?

In order to make the olf and decipol methodology work there would be a need for data on materials and emissions from materials. This is also true of, for example, the TVOC method. Furthermore control of materials is a good thing in itself. In fact a considerable effort is under way on emissions from building materials, both within CEN and under the European Collaborative Action: Indoor air Quality and its Impact on Man. Work is under way in CEN TC112 on formaldehyde emissions from wood based products and in CEN TC264 on chemical emissions from building materials. An interlaboratory comparison of test chambers has already been carried out under the European Collaborative Action: Indoor air Quality and its Impact on Man, and a second interlaboratory comparison is underway. CEN TC264 is due to produce a draft standard for public enquiry in 1997 for publication in 1998.

Another approach which has considerable potential is the so called electronic nose, which in fact consists of a number of conducting polymer receptors which are doped to make them respond to desired substances. If agreement can be reached on the substances to be detected and the weighting given to these substances, we may have an objective and repeatable decipol. Still problems will remain relating such a measurement to peoples' dissatisfaction but at least the basis will be a physical measurement.

Conclusion

A great deal of work needs to be done, particularly on materials, which would be needed, whatever approach to the problem is finally adopted. This does not mean that the production of the document has been a waste of time. Rather perhaps it draws attention to the difficulty of trying to draw up a standard whilst research is ongoing. The CEN report which seems likely to emerge will represent part of the state of current research.

References

1. Oseland N A and Raw G. "Perceived Air Quality ; Discussion of the new units". BSERT Vol 14 No 4, 1993, CIBSE London
2. Knudsen H N, Clausen G and Fanger P O. "Characteristics of sensory emission rates from materials". Healthy Buildings 1994, Budapest, Sept 1994.

CHAPTER SEVEN

An Adaptive Approach to Thermal Comfort Criteria

Michael Humphreys

Research Consultant, Dunstable, UK

Abstract

People are not passive receptors of their thermal environment. They take actions in order to improve their thermal comfort. These actions or adaptations include modifying the rate of internal body heat generation, modifying the rate of body heat loss, modifying the thermal environment, and selecting a different environment. The constraints which act on these adaptive processes result in the particular temperatures for thermal comfort for that person. The climate acts as an overarching constraint, and may be used, in conjunction with the results of thermal comfort field-studies, to suggest temperatures appropriate for human occupation. A well designed and well operated naturally ventilated building in the UK need not, on these criteria, be uncomfortably hot in summertime.

Keywords: Adaptive model, thermal comfort, climate, natural ventilation.

1 Introduction

The usual approach to understanding thermal comfort is the analysis of the heat exchange between a person and the environment. The best known model is set out in Fanger's book 'Thermal Comfort' [1] and expressed by his Comfort Equation, which has been adopted by ISO as Standard 7730 [2]. On this model discomfort is attributed to an imbalance between a person's heat gains and heat losses. A simple application of ISO 7730 frequently leads to an incorrect evaluation of thermal discomfort [3, 4]. The discrepancy becomes large in environments which are very different from the conditions under which the model was calibrated. It over-estimates the heat stress of a warm indoor climates, and the cold stress of a cool indoor climate.

Naturally Ventilated Buildings: Buildings for the senses, the economy and society. Edited by D. Clements-Croome. Published in 1997 by E & FN Spon. ISBN 0 419 21520 4

The reasons for the discrepancies are not yet fully understood, but appear to be attributable to an inadequate allowance for people's physiological, psychological and behavioural adaptive responses to the indoor and outdoor climates. An alternative approach, avoiding the use of ISO 7730, is therefore advisable. Such an approach is possible by using an adaptive understanding of thermal comfort.

The Adaptive Model does not start from heat exchange. Rather it notices that people take a whole range of actions to secure satisfactory conditions, and makes this adaptability the central point for understanding thermal comfort, and for predicting discomfort. We may state an Adaptive Principle:

> *If a change occurs such as to produce discomfort,*
> *people react in ways which tend to restore their comfort.*

A person is not a passive receiver of sense-impressions, but is in dynamic equilibrium with the thermal environment. Thermal pleasure is usually associated with the return towards thermal neutrality, and is dynamic rather than static in character. The Adaptive Model does not in principle conflict with the heat exchange model, for adjustments to the heat exchange process may be among the actions taken in order to secure comfort.

2 Adaptations

The principal seat of body temperature regulation is located in the brain. Its temperature is closely controlled, and should the brain temperature deviate from normal, actions are initiated which tend to return it to normal. Below is a list of some of the actions which the brain may initiate if it becomes too cold. If the actions fail, the result is discomfort, and possibly hypothermia and death.

> Vasoconstriction (reduces blood flow to the surface tissues)
> Increasing muscle tension (generates heat in the muscles)
> Shivering (generates more heat in the muscles)
> Curling up (reducing the surface area available for heat loss)
> Jumping about or otherwise increasing activity (generates body heat)
> Eating a meal (digestion produces heat in the body)
> Cuddling up (reduces surface area available for heat loss)
> Adding clothing (reduces the rate of heat loss per unit area)
> Turning up the thermostat (perhaps raises the room temperature)
> Lighting a fire (usually raises the room temperature)
> Finding a warmer spot in the house (selects a warmer temperature)
> Visiting a friend (hoping for a warmer temperature)
> Chopping some wood (increases body heat and may lead to a warmer room)
> Insulating the loft (long term hope of warming the house)
> Double glaze the windows (hope to raise temperatures and reduce draughts)
> Insulate the wall-cavities (hope to raise the indoor temperatures)
> Fill cracks and draught-strip the doors and windows (hope to reduce draughts and raise room temperatures)
> Claim a heating allowance (to buy fuel and raise indoor temperature)
> Build a new house (planning to have a warmer room temperature)

Emigrate (seeking a warmer place long-term)
Go to bed (seeking a warmer place short-term)
Go to the library (hoping for a warmer room temperature)
Complain to the management (hoping someone else will raise the temperature)
Put a hot water bottle up the jumper (apply heat locally)
Acclimatise (let your body and mind become more resistant to cold stress)

If the brain gets too warm it may initiate some of the actions shown below (some of these do not work in very hot environments). If the actions fail there will be discomfort and perhaps death from heat-stroke.

Sweating (loses heat by latent heat of evaporation)
Vasodilatation (increases blood flow to surface tissues)
Adopt an open posture (increases the area available for heat loss)
Stop cuddling up (increases the area available for heat loss)
Strip off some clothing (increases heat loss per unit surface area)
Slow down activities (reduces bodily heat production)
Go for a swim (selects a cooler environment)
Have a beer (induces sweating, thus increasing heat loss)
Eat less (reduces body heat production)
Have a cup of tea (induces sweating, more than compensating for its heat)
Adopt the siesta routine (match the activity to the thermal environment)
Find a cool spot (selecting a cooler environment)
Turn on the air-conditioning (lowers the air-temperature)
Turn on a fan (increases air movement, which increases heat loss)
Sit in a breeze (selecting a higher air-movement)
Open a window (brings room temperature nearer outdoor temperature, which is often cooler - can also increase the air-movement)
Visit a friend (hoping to select a cooler temperature)
Build a better building (long term way of finding a cooler spot)
Emigrate (long-term way of finding a cooler place)
Acclimatise (let the body and mind adjust so that heat is less stressful)

The various actions may be classified into four types:

Modifying the rate of internal heat generation
Modifying the rate of body heat loss
Modifying the thermal environment
Selecting a different thermal environment

If all the avenues of adaptation were freely available, it would not be possible to predict which way people would use to get comfortable, and so we should be unable to predict what room temperatures they would find comfortable or uncomfortable. From the point of view of building design and operation, the comfort temperature may be regarded as the 'answer' or 'result' or 'solution' of the adaptive processes. It is frequently this temperature which we wish to know for design purposes. However, because not all options are always available, a knowledge of the constraints which are operating can lead to a knowledge of the solutions (comfort temperatures) which are applicable.

3 Constraints

Scrutiny of the lists of adaptive actions shows that some may not be practicable in the prevailing circumstances. There are usually certain *constraints* in operation. Constraints may arise from climate, fashion, culture, religion, conscience, poverty, building design, or social context. *Constraint* is being used in a quasi-technical sense. Constraints may alter the comfort temperature, and may prevent comfort being achieved. We note and comment on some common constraints. The list is illustrative rather than exhaustive.

Climate Usually there is little a person can do about the climate and the weather, as few can go abroad whenever they wish, or choose their place of work according to climate. Climate is therefore an overarching constraint, affecting the comfort temperatures and sometimes preventing comfort being achieved.

Short term and long term changes A glance at the lists of adaptive actions shows that the times needed to accomplish them vary enormously. Emigration or building a better house, office or factory may take more than a year to plan and achieve. Filling a hot water bottle is a matter of minutes. It follows that there are more ways available to respond to a gradual change of the environment than there are to a sudden one. Therefore sudden change is a stronger constraint than is a gradual change, and more likely to cause discomfort.

Cost Some strategies are free, like curling up. Others are costly, like re-building. The costs of many actions would be prohibitive to the poor. Therefore poverty introduces constraint, and the poor are likely to suffer more discomfort than are the rich. On a larger scale the company's and the nation's economies are constraints.

Relative costs Strategies may have different relative costs in different countries. Clothes may be relatively cheap and fuel relatively costly, or the other way round. Therefore different populations will in general settle for different comfort temperatures because of the relative strengths of particular constraints, even if the general level of constraint is equal in the different populations.

Personality A person with a rigid personality will be operating under personally imposed constraints which would not affect a person with a more flexible and ingenious approach to living. The latter will devise new strategies while the former will exclude some available strategies. A rigid personality is a constraint, and therefore rigid people are prone to discomfort ... though perhaps more prone to grin and bear it? The effect may extend to the cultural ethos of the company or nation.

Gender In many cultures the clothing of men differs from that of women, and the constraints imposed upon men and women also differ. In general therefore one would expect the sexes to have a different proneness to discomfort, and perhaps different comfort temperatures.

Physiology The body's temperature regulation system is not fully developed in the new-born child, and deteriorates in old age. This is potentially dangerous because discomfort may not be manifest, and the requisite life-saving actions may not be taken. Babies and the aged may be at risk from cold and heat, because the adaptive principle operates less powerfully for them. This dangerous condition is usually overcome culturally, babies and the elderly sharing the same accommodation as people who have adequate body temperature control. There is danger in some societies where babies sleep apart from their mothers and where the elderly live alone.

Fixed glazing This is a constraint if it reduces the occupant's control over the thermal environment. It may both alter the comfort temperature and cause discomfort.

Uniforms A uniform is a constraint on the free choice of suitable clothes, and is therefore likely to alter the comfort temperature and cause discomfort. This effect is lessened if the uniform consists of a number of accepted options which the person may freely choose.

Fashion In some sections of society fashion may amount almost to a uniform, and a person may choose discomfort rather than be out of fashion. It should be regarded as a constraint. We note the possibility of a trade-off between thermal comfort and 'social comfort'.

Formal occasions There are fashions for certain social, academic, ceremonial or work situations, which constrain the free choice of clothing. Formality is a constraint, and therefore likely to alter comfort temperatures and produce discomfort.

Fixed locations Restricting a person's movement is a constraint. Examples include the supermarket check-out person, the assembly-line worker, and the bank cashier. Fixed locations are likely to produce discomfort.

Thermal control operated by another Where ever the control is given to someone or something other than the person whose comfort we are considering, there is constraint. Examples include going to the theatre, going out for a meal, being at the workplace if the environment is controlled by the management or its nominees. These variations are imposed rather than chosen, and therefore likely to produce discomfort rather than assist comfort. Whenever a group of people together occupy the same thermal environment there is constraint of this kind, for the individuals have lost control over their environment.

The requirement to save energy The requirement to save energy may restrict the temperature settings available to the occupant. This may constrain the adjustment of the temperature as a means of restoring comfort. So energy conservation can lead to discomfort, unless it is thoughtfully and appropriately undertaken. There may of course be an acceptable trade-off between cost saving and comfort, if the occupant shares in the benefit from the saving.

Some actions which help to restore comfort may be ruled out because of social, cultural, commercial, fashionable, conscientious, religious or economic constraints. This comment is particularly (but not exclusively) true of variations in clothing, for clothing has many functions besides thermal insulation and is subject to wide variations from culture and fashion. Those investigating thermal comfort, or seeking to provide it, must always be aware of possible constraints, for they indicate a likelihood of discomfort. An example of maximum constraint is a soldier on parade. The soldier has no freedom of action, and the comfort-temperature is therefore completely fixed. If the temperature differs appreciably from the comfort temperature the result is discomfort or worse. An example of minimum constraint is a wealthy person alone in their own home. There is almost total freedom, the comfort temperature is highly adjustable, and so is the room temperature. So there is no reason for a mismatch, and no reason for discomfort.

4 Designing and building for comfort

From the above understanding of dynamic adaptation we can see that the building designer can do much to make a building comfortable for the occupants.

The environment should be *predictable* - the potential occupant should know what to expect, so that suitable clothes can be chosen. In practice this means that temperatures should be reliably the same in similar accommodation on the various occasions when people visit them, either for work or leisure. This need not exclude a seasonal variation, if this is expected by the occupants and their visitors.

The environment should be *normal* - that is to say, it should be within the range acceptable in the social circumstances prevailing in that society and climate. Subgroups may of course develop their own norms; some prefer to keep their environment warmer than others. One must bear in mind that what is normal for the subgroup may be unpredictable for a visitor.

Where people are free to choose their location, it helps if there is plenty of usable thermal *variety*. Then they can choose places they like, which are suitable for the activity in which they wish to engage. (Watch a cat finding the best place for a sleep.) The same principle applies to the design of outdoor spaces, courtyards, parks and gardens. The provision of sunny and shady spaces, sheltered and open routes, enhances comfort by giving freedom of choice.

Where people must be at a fixed location, we should provide them with *adequate control* of their thermal environment.

Where people are in uniform or their dress is otherwise constrained, we should provide them with adequate *individual control* over their thermal environments.

We should avoid sudden *imposed* changes of temperature. This means that one should specify thermostats with small switching intervals, or, in naturally ventilated designs, buildings with sufficient interior thermal capacity. At present there is not enough knowledge to tell how rapid or how big the changes may be, and this is a subject currently receiving further research. We should distinguish between imposed variation and chosen variation. Imposed variation is likely to produce discomfort, while chosen variation is likely to reduce discomfort. This is because the latter is correlated with the desires of the occupant, while the former bears no relation to those desires. It follows that exactly the same temperature variation may be acceptable or unacceptable, depending on whether it is chosen or imposed.

All this suggests that a naturally ventilated heavy-weight building, whose interior temperature in summer drifts slowly in sympathy with the variation of the seasonal temperature, should be capable of providing a comfortable thermal environment for its occupants. In the British climate overheating should not become a problem, even in the hottest weather, unless the building is subject to high internal heat loads.

5 Energy-conscious design and thermal comfort

Minimising energy use may entail the following strategies (the list is not exhaustive.):-

> Heating or cooling the person rather than the space
> Using ceiling fans rather than refrigeration
> Using the building envelope to provide suitable environments rather than
> relying on heating or cooling plant
> Designing in sympathy with the climate

Allowing seasonal drifts in indoor temperature within prudent limits
Using incidental heat gains in winter; extracting heat at source in summer

It is important to notice that these measures may impose constraints on the occupants, and produce discomfort. It is therefore desirable to give effective control to the occupant, or the level of constraint may be excessive. Heated or cooled work-stations should be controlled by the occupant, so should fans, blinds, etc. There is no objection to sophisticated ('intelligent') control systems, so long as they are under the effective control of the user - like having a manual over-ride on an automatic gearbox. But we often over-estimate how well people understand the controls they are given. For example, few people in Britain understand that closing windows on a hot day can contribute to thermal comfort, while this procedure is a matter of course in Southern Europe.

6 How powerful are the adaptive processes?

6.1 Mean temperatures and comfort temperatures
If adaptation were in general successful, we would expect that on average, the conditions people have would be the conditions they find comfortable. That is to say, the comfort temperatures ought approximately to equal the mean temperatures of people's experience. The results from numerous field-studies of thermal comfort world-wide over many years shows that the relation is very close [5].

6.2 Damping down of the effects of changing temperature
We would expect people's reaction to varying temperatures to be less than that obtained in the laboratory, where adaptation has been eliminated by standardising the clothing and the activity. Table 1 compares the sensitivity to temperature change found in laboratory studies [1] with those common among field studies [5]. The sensitivity is measured in comfort-scale units per deg C, on a seven-point warmth scale such as the ASHRAE scale or the Bedford scale. The table shows that the degree of adaptation depends on the time available for the adaptive processes to be completed. The adaptation to seasonal change of temperature is virtually complete.

Table 1 Comparison of sensitivities to temperature changes in different circumstances:

Climate chamber experiments, standard clothing and activity	0.32	scale units per degree
Field study, observations extending over some days	0.23	
Field study, once per day, over a whole year	0.16	
field study, once per week over a whole year	0.10	
Field study, monthly means of responses over 15 months	0.05	
Field studies, between studies regression coefficient	0.05	

6.3 Evidence of a progressive approach to equilibrium
If comfort data are obtained sequentially over long periods it is possible to calculate a time-constant for the adaptive processes, rather like the concept of the half-life for

radioactive decay. Evidence so far available indicates that many adaptations become complete within a few days, but this is an area where more information would be welcome [6]. Further research is now in progress.

7 What governs the temperatures found to be comfortable?

The comfort temperatures resulting from the adaptation processes are related to the climate, to the society to which the data apply, and to whether a building is consuming energy for space heating and cooling [7]. As far as climate is concerned, the best correlate so far discovered is the prevailing mean outdoor air temperature, which may be found from meteorological tables. There may also be an effect arising from the maximum temperatures in the summer. As far as society is concerned, North Americans on the whole have adopted higher winter indoor temperatures than have Europeans, climate for climate. This is another area where more detailed knowledge would help the designer. Standards can operate as self-fulfilling prophecy - if 25 deg C is specified and provided, then people will adapt to it, and report it as comfortable. If 22 deg C were specified and provided, the same would apply.

Figure 1 shows dependence of comfort temperatures on monthly mean outdoor temperatures and upon the mode of operating the building. Table 2 sets out the percentages of the variation of the comfort temperatures world wide which are accounted for by the factors listed above. Already from this knowledge it is possible to predict comfort temperatures more accurately that by using heat exchange theory.

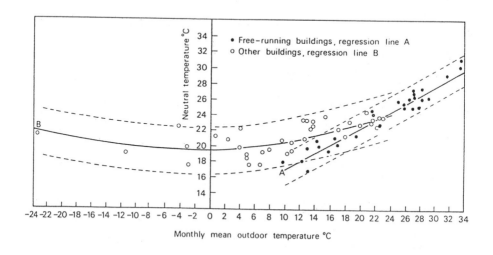

Fig. 1. Scatter diagram for neutral temperatures (reproduced from [6])

The comfort temperature in a free running building can be estimated with a standard error of 1 deg C. For heated and cooled buildings, with a standard error of 1.4 deg C.

Table 2 Variance of comfort temperature explained by the adaptive model

a) Free running buildings:	Monthly mean outdoor temperature	95%
b) Heating or cooling plant in use:	Monthly mean outdoor temperature	44%
	Plus mean daily max. of the hottest month	59%

It seems to follow that if a naturally ventilated building can be designed and operated so that its temperature approximates to that indicated on the line for 'free-running' buildings, the occupants will find it comfortable. That this is not in principle difficult to achieve in climates as temperate as that of the UK is evident from the fact that many of the quite ordinary buildings in which comfort surveys have been conducted were satisfactory in the mean temperatures they provided. The common failing was too great a diurnal swing in indoor temperature, which was attributable to too light a type of construction.

An experimental algorithm has recently been suggested which could be used to evaluate the thermal performance of a naturally ventilated building in summertime. It relates the indoor temperature to a running mean of the prevailing outdoor temperature [8].

8 Acknowledgements

This paper is a revision of one prepared for the Workplace Comfort Forum, RIBA, March 1995. Figure 1 is reproduced by permission of the Controller of HMSO: Crown Copyright 1995.

9 References

1. .   Fanger, P.O. (1970) *Thermal Comfort, Danish* Technical Press, Copenhagen.
2.   International Standard 7730, (1984), ISO Geneva, revised 1990.
3.   Humphreys, M.A. (1992) Thermal Comfort Requirements, Climate and Energy. *Proceedings of the 2nd World Renewable Energy Congress,* Volume 4, Ed. Sayigh, A.A.M., Pergamon.
4.   Humphreys, M.A. (1994) Field studies and climate chamber experiments in thermal comfort research. In: Thermal comfort: past, present and future. Eds N.A. Oseland and M.A. Humphreys, Building Research Establishment Report , Watford.
5.   Humphreys, M.A. (1975) Field Studies of Thermal Comfort Compared and Applied. Building Research Establishment Current Paper CP 76/75.
6.   Humphreys, M.A. (1978) The Influence of Season and Ambient Temperature on Human Clothing Behaviour. In: Indoor *Climate,* Eds. Fanger, P.O. and Valbjorn, O., Danish Building Research Institute, Copenhagen.
7.   Humphreys, M.A. (1978) Outdoor Temperatures and Comfort Indoors *Building Research and Practice* 6(2) 92-105.
8.   Humphreys, M.A. and Nicol, J.F. (1995) An adaptive guideline for UK office temperatures. In: *Standards for Thermal Comfort,* Eds: F Nicol, M Humphreys, O Sykes, & S Roaf. E & FN Spon, London

CHAPTER EIGHT

Natural Ventilation: Prediction, Measurement and Design

Phil J Jones
Welsh School of Architecture, UWCC, Cardiff, UK

Abstract

This paper reviews some of the current procedures available for predicting and measuring ventilation performance of for naturally ventilated buildings, and in particular how measurements and prediction methods can be combined to provide a better understanding of ventilation performance. The paper includes a brief review of :

- air leakage measurements (whole building and component testing)using fan pressurisation;
- thermography to locate main air leakages;
- measurement of ventilation rate using tracer gas and metabolic carbon dioxide;
- visualisation and measurement of indoor air movement;
- prediction of ventilation rate using zonal models;
- prediction of indoor air movement and ventilation rate using CFD;
- wind tunnel estimation of external pressure coefficients (Cp's).

Throughout the paper reference will be made to results from recent research projects carried out by the Architectural Science Research Group at the Welsh School of Architecture.

Naturally Ventilated Buildings: Buildings for the senses, the economy and society. Edited by D. Clements-Croome. Published in 1997 by E & FN Spon. ISBN 0 419 21520 4

1 Introduction

The ventilation system in a building must be designed :

- to maintain minimum fresh air ventilation appropriate to the needs of the function of the space in order to achieve good air quality;
- to avoid excessive ventilation during the heating season to avoid an energy penalty;
- to deliver fresh air to the occupied space in a manner that does not give rise to discomfort through draughts or low air temperatures.

The prediction of ventilation performance should be included :

- at the concept design stage - ventilation design will impact on building form and spatial planning ;
- at the detailed design stage - to position and size openings and other ventilation devices.

The measurement of ventilation performance should be included :

- at the completion of construction - in order to assess performance in practice, either as part of the 'hand-over' quality assurance procedure or as a research activity to improve understanding of ventilation design.

The main concepts associated with natural ventilation design are becoming better understood. However, the guidelines and procedures for the detailed design of a natural ventilation system are not clearly defined, and relatively little is known about actual ventilation performance in practice. There is insufficient information on how natural ventilation varies over time and how it performs spatialy, for example in relation to depth and volume of space. Recently there has been an increase in buildings that have some form of 'hybrid' natural/mechanical ventilation system, again for which there are no established technical guidelines. There is therefore the need to assess what prediction techniques are available for informing ventilation design and for measuring ventilation rates in practice, and, to what extent these available for use by designers.

This paper addresses two areas :

- the main prediction techniques that are available for design for natural ventilation;
- the measurement techniques that are available for assessment of ventilation performance in practice.

2 Ventilation Performance Assessment by Measurement

There are three types of measurement to assess ventilation performance :

- measurement of air leakage through the fabric, providing a measure of infiltration rate;
- measurement of actual ventilation rates, combining infiltration and purpose ventilation (through windows, vents, etc.);
- measurement of indoor air movement resulting from the combination of ventilation and internal sources of heat and momentum, to relate to air quality and comfort.

2.1 Air Leakage

2.1.1 Whole Building Air Leakage

The air leakage of a building can be measured by fan pressurisation. This provides a measure of overall air leakage which includes leakage through the construction and its details as well as the leakage associated with specific components, for example, around closed doors, closed windows and service entries. The envelope leakage is measured over a specified range of internal/external pressure differences, usually up to 50-60Pa, which is above the influence of pressures induced by the prevailing wind and temperature differences. A fan is used to either pressurise or depressurise the building. In order to ensure that prevailing weather effects are minimised the measurement should be carried out when wind speeds are less than 2m/s. The internal/external pressure difference is usually achieved by locating the fan in a door. Figure 1a/b shows the equipment in use for single and multiple fan pressurisation measurement. The technique is relatively easy to carry out on domestic scale buildings where small fans can produce the necessary flow rates to achieve the required pressure difference. In larger buildings, either multiple fan (as shown in Figure 1b) or larger single fan systems are needed. Because of the practical limitations in fan capacity, it is more difficult to achieve the required pressure differences in larger buildings, and often lower inside/outside pressure differences are used (say up to 30Pa) and the results extrapolated to 50Pa.

Fig 1 : Fan pressurisation equipment for domestic (left) and non-domestic (right) application.

2.1.2 Component Air Leakage

Fan pressurisation cannot be used to locate air leakage. However, the air leakage associated with specific components of a building can be measured either by reductive sealing or component testing methods. In the case of the reductive sealing method, the air leakage of the whole building is measured and then the element of interest is sealed. The building is pressurised again, and the difference between the unsealed and sealed test provides a measure of air leakage associated with the sealed element. Figure 2 shows the sealing of a factory loading door and Figure 3 shows the corresponding reduction in leakage for the sealed and unsealed measurements.

Fig 2 : The 'leaky' roller shutter door of a small factory unit is being sealed to assess its contribution to the overall air leakage.

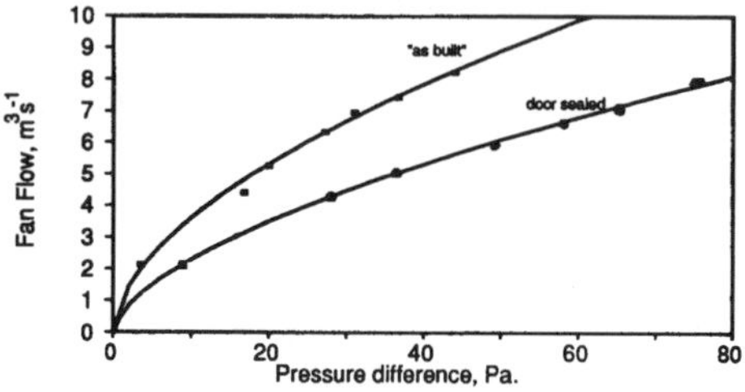

Fig 3 : The air leakage curves for a factory with and without the loading door sealed [1].

Tests have shown that the specification of 'well sealed' components such as doors, window frames, vents, etc., can considerably reduce overall air leakage, and thus result in lower ventilation rates and energy use. In some industrial units air leakage can easily be reduced by 50% or more by using well sealed components, as indicated in the experimental results in Figure 4.

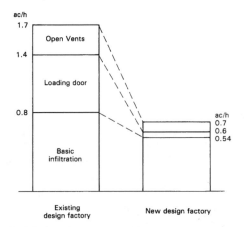

Figure 4 : Reduction in air leakage between an existing and new design factory due to the specification of better sealed components and reduced infiltration through the fabric [2].

Component testing involves pressurising individual components or element of a building by containing them in a temporary enclosure and directly pressurising the component.

2.1.3 Identifying Air Leakage Using Thermography

Fan pressurisation is useful for quantifying air leakage. However, because it does not identify the location of the leakage, its use in practice is limited to providing an overall comparative measure of air leakage as a quality assessment indicator. A thermography survey, although not able to quantify air leakage, can however be used to locate it, by identifying local heating or cooling of the fabric in the leakage area. Figure 5 shows how thermographic images can be used to detect air leakage on the inside or outside of the building envelope [3].

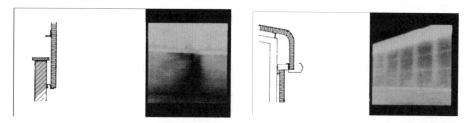

Figure 5 : Thermal image of local cooling due to cold air infiltration at the internal cladding masonry wall detail (left) and local heating due to warm air exfiltration through the external eaves detail (right)

In order for air leakage to take place there has to be a pressure difference between inside and outside. Usually there is sufficient pressure difference generated from an inside/outside air temperature difference of about 10°C to identify the major leakage areas. This is the standard inside/outside temperature difference needed to carry out a thermographic survey for the more usual application of assessing installation of insulation. The thermal image will be enhanced by combining thermography with fan pressurisation, as indicated in Figure 6 below, although fan pressurisation is not usually necessary to reveal leakage locations.

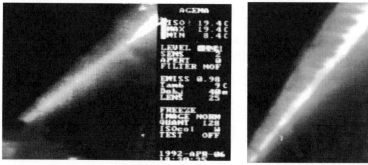

Figure 6 : Thermal image of an external eaves detail, without (left) and with (right) accompanying fan pressurisation.

For fan pressurisation air leakage tests and thermographic surveys, there are a number of specialist contractors that offer a service. Although the thermographic measurement technique is not difficult to understand and the equipment is relatively easy to use, a level of experience is necessary for interpreting the thermographic images in relation to the identification of air leakage as different from the image due to defective insulation.

2.2 Ventilation Rate Measurement Using Tracer Gas

There are a number of methods available for measuring ventilation rate involving the injection and measurement of a tracer gas in a space. The discussion here will concentrate on the most common and accessible methods.

The injection and measurement of tracer gases in a space can provide a direct measure of the ventilation rate. There are three main tracer gas measurement methods in use, namely the 'tracer decay', the 'constant concentration' and the 'constant emission' methods, all using derivatives of the continuity equation below.

$$V \frac{dC}{dt} \quad = \quad F \quad + \quad QCe \quad - \quad QCt$$

| Change in tracer concentration in space | Amount of tracer introduced into space | amount of tracer leaving space |

Where :   V is the volume of the space (m³),
           F is internal rate of production of tracer gas (m³/h),
           Q is volume flow of air into the space from outside (m³/h),
           t is the time (h),
           C,Ce are the internal and external levels of concentration of tracer gas at time t (m³/m³).

2.2.1 Tracer Decay

This method consists of injecting tracer gas into the space to achieve a uniform target concentration. Once the target concentration has been achieved all injection of tracer gas is stopped. The tracer concentration is then allowed to decay. Under perfect mixing the tracer concentration would decay exponentially and would be described by the expression.

$$Ct = Co \exp(-(Q/V)t)$$

The logarithmic version of this equation describes a straight line graph of the form : $\ln Ct = \ln Co - (Q/V)$, where theair change rate (Q/V) is then the slope of the graph.

Figure 7 shows a typical decay curve for a large space. For larger spaces it may be necessary to sample at more than one point.

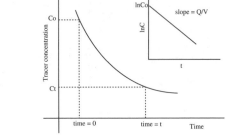

Figure 7 : An example of a tracer decay curve

Tracer gas tests can use injected tracer gases such as nitrous oxide, SF_6. Alternatively tests can use the decay of metabolic carbon dioxide after occupancy , provided the space is emptied of all occupants over a short period of time. Bag sampling techniques can be used to both inject and sample tracer gas and the sample gas concentrations collected in the air bags are analysed remotely.

2.2.2 Constant Concentration Methods

Continuous measurement of ventilation rate can be carried out using constant concentration methods, which provide a measure of the variation of ventilation rate over time, for example, due to changes in internal and external conditions.

The method involves the injection and sampling of a tracer gas to maintain a constant concentration of the gas in the space. Fans can be used to maintain good mixing of the tracer gas in the space and generally the tracer gas is sampled and injected at a number of points throughout the space (for large spaces) or in each room (in domestic scale buildings). The continuity equation is of the form :

$$Q/V = F/(V.C)$$

Where (Q/V) is the ventilation rate. The ventilation rate is directly proportional to the tracer gas injected for a constant concentration.

The constant concentration method requires computer controlled injection and sampling units which are relatively expensive and complex in their calibration and setting up. The method therefore is more difficult and time consuming to carry out than the tracer decay method. Figure 8 shows the equipment needed to carry out a constant concentration test. Figure 9 shows typical results [4].

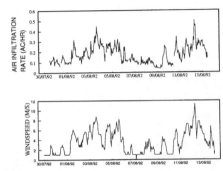

Figure 8 : Constant concentration tracer gas equipment using N_2O as a tracer gas.

Figure 9 Variation in ventilation rate (top) in relation to wind speed (bottom) for a 900m^2 factory unit.

2.2.3 Constant Emission Rate

For occupied spaces it is not usually practicable to carry out tests which require complex equipment and 'unfamiliar' tracer gas. In such cases metabolic carbon dioxide can be used as a tracer. Using assumed emission rates, based on observed occupancy levels, combined with measured concentration levels, the ventilation rate can be estimated using a constant emission analysis.

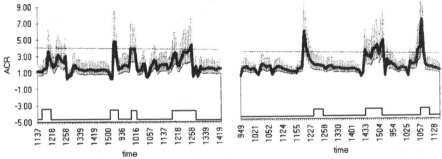

Figure 10 : Ventilation rates estimated from constant emission of CO_2 from occupants to assess variation in ventilation in a school classroom. Window opening is indicated on the horizontal axis [5].

2.3 Visualisation of Indoor Air Movement

The pattern of indoor air movement will determine the effectiveness of heat distribution and fresh air ventilation, which will relate to indoor air quality and comfort. Indoor air movement can be visualised and quantified by experiment.

Internal air flows can be visualised using some form of seeding method, with smoke, balloons or bubbles, or some other neutrally buoyant particulate. Visualisation involves three stages, seeding the air, illuminating the seed and recording the visualised flow.

Smoke can be used as a qualitative means of measuring air movement (Figure 11). It can be injected into the space using a smoke generator and then observed and recorded using video or manual (sketching) methods. Smoke puffers are also available for tracing local air movement. These are small hand-held devices that emit smoke and are particularly useful for locating and demonstrating component leakage (eg. draughts around a window). It is often not appropriate to use smoke in sensitive spaces.

Neutrally buoyant balloons or bubbles can also be used in a qualitative way to assess space air movement in a similar way to smoke. Balloons in particular have the advantage that they are relatively clean and can be used in sensitive spaces, eg. museums. For a more quantitative measure of air movement, their position in space can be monitored with multiple video cameras [6] The video held information can then be computer analysed to provide accurate three dimensional velocities and streamlines of air flow in the space, eg. patterns of air movement in an atrium.

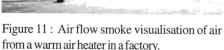

Figure 11 : Air flow smoke visualisation of air from a warm air heater in a factory.

Figure 12 : Multiple images taken from a video recording of a neutrally buoyant balloon indicating the predominant air flow pattern.

3 Ventilation Rates Predicted

There are a range of ventilation prediction methods available. This paper will concentrate on the two more complex and emerging methods, namely zonal models and CFD.

3.1 Zonal Models

Zonal or network models can be used to calculate the flow of air between one or more zones and the outside. Their main advantage is that they can calculate inter-zone flows and therefore air change rates, ventilation heat transfer and the transfer of contaminants. They can be used to study new building forms. They can handle a wide range of opening and crack types and can predict the interaction of buoyancy and wind driven effects. They are computer based and calculate the flows between pressure nodes within the building and to outside. One example of a zonal model is HTB2-VENT [4], which can be used to predict interzonal flows, separate or as part of a dynamic thermal prediction.

In order to set up a zonal model prediction the main air leakage locations have to be identified. Figure 13 shows the main air leakage locations for a single space industrial unit.

Figure 13 : Location of major sources of air leakage, from construction details and components, over the building envelop.

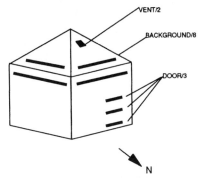

To assess the validity of zonal models, leakage data from pressurisation tests, with leakage sites located by thermography, can be used as input data by the zonal model to predict variation in ventilation rate over time. Comparison can be made with measured ventilation rates from constant concentration tests as shown in Figure 14 [4]. Comparison of measured and predicted rates can be in close agreement, provided the input data on leakage is accurate.

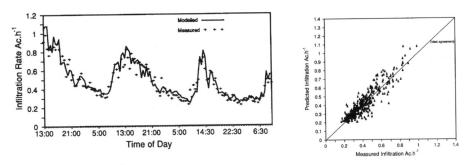

Figure 14 : Comparison of predicted and measured ventilation rates over time (left) and as an x-y plot (right).

Once confidence has been established with the zonal predictive model, it can be used to explore what would happen, for example, if the leakage area of a building was halved or doubled, as shown in Figure 15. The zonal model can be combined with an energy model to predict seasonal ventilation heat loss as shown in Figure 16 [4].

Figure 15 : Variation in ventilation rate with leakage area halved (a tight building) and doubled (a leaky building) compared to a normal level of leakage.

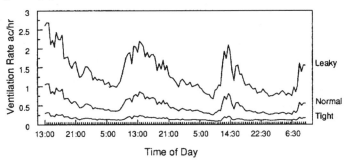

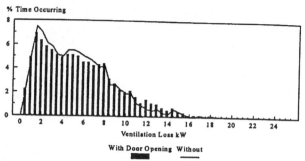

Figure 16: The application of a zonal ventilation prediction model with a dynamic thermal model (HTB2) to predict seasonal variation in ventilation heat load for a factory with and without the influence of loading door opening.

3.2 CFD Prediction

Computational fluid dynamics (CFD) can be used to predict internal air movement and ventilation. It can predict the air movement and temperature distribution arising from sources of momentum (ie. jets), surface heat transfer and pressure boundaries (ie. natural ventilation) and can acount for the effects of blockages due to the geometry of the space and its contents. It can also account for the interaction of pressure and buoyancy forces (Figure 17).

CFD can predict 3-dimensional effects that are either constant or variable over time. It is therefore an extremely versatile and useful technique in the field of ventilation prediction. It is however highly complex and requires a level of skill and understanding of ventilation design, building physics and computational numerical techniques in order to obtain credible solutions. Models are becoming easier to use by the non-specialist and the need to use such models in ventilation design will eventually result in their widespread acceptance. In the meantime there are specialist consultants in the area who will undertake ventilation simulation work.

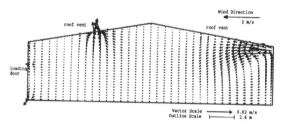

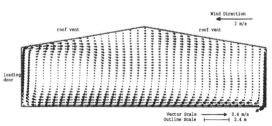

Figure 17: Internal air movement due to wind induced pressure boundary conditions at leakage around details (top), and combined with internal buoyancy effects (bottom). Internal buoyancy dominates the pattern of air movement[4].

CFD is particularly useful for predicting air flow paths in naturally ventilated spaces where the building form and space planning are an integral part of the ventilation design (Figure 18). Comparison with measured data has indicated that CFD predictions can be sufficiently accurate to inform the design process [7,8].

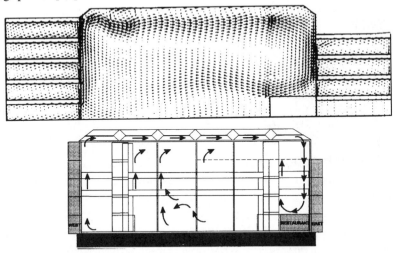

Figure 18 : CFD prediction of internal air flow in a naturally ventilated atrium with perimeter offices (top). Results from balloon visualisation are in close agreement with CFD predictions (bottom) .

3.3 Wind Tunnel Modelling

Figure 20 :　Boundary layer wind tunnel

In order to predict ventilation due to wind effects the pressure distribution over the external envelope of the building must be estimated. A physical model of a building and its surroundings can be constructed and placed in a wind tunnel (Figure 20) where it is subjected to a controlled wind flow. Pressure sensor taps can be installed at various points on the building envelop, corresponding to ventilation openings. The pressure at each opening can be measured using a pressure sensing manometer. This can then be related to the free wind pressure, at a point of known height above the surface, in order to obtain the Cp value. Figure 21 illustrates the use of pressure taps on a cube model of a building and the resulting pressure distribution. The main use of wind tunnels for ventilation studies is to predict Cp's. However, they can also be used for flow visualisation by introducing smoke or other tracers in the wind tunnel and observing the flow characteristcs. . Most building application wind tunnels operate at scales of between 1:100 to 1:500. As long as the Reynolds Number is kept high (through high tunnel air speeds - usually about 5 to 10 m/s) the turbulent regime is ensured and scaled and real flows will match.

Values of Cp measured in the wind tunnel can be used as pressure boundary input data to simple or more comlex zonal and CFD ventilation models.

 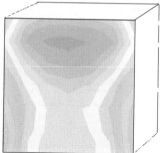

Figure 21 : Pressure tappings in a model of a 'cube' building (left) and the resulting surface presssure contours (right) [9].

4 Conclusion

This paper has identified a number of prediction and measurement techniques that have been used in natural ventilation design, and research. Where measurement and prediction results have been compared there has been good agreement. Most of the techniques described need specialists with specific skills to apply them to design problems. These skills should be developed within the design profession together with guidelines and procedures in the application of ventilation models and measurements

Acknowledgements : The work described in this paper was carried out in the Architectural Science Research Group and the author acknowledges the contributions from colleagues - Don Alexander, Greg Powel and Hugh Jenkins.

References
1 P J Jones and G Powell - Comparison of Air Infiltration Rate and Air Leakage Tests Under Reductive Sealing for an Industrial Building, *10th AIVC Conference Finland,* (1989) 131-152 ISBN 0 -946075-468.
2 Insulated Factory Loading Doors, Expanded Project Profile 334 Energy Efficiency Office (1988).
3 P J Jones and G Powell, An Investigation of Insulated Cladding Constructions for Industrial Buildings, *Final Report to WDA*, (1993).
4 P J Jones, D K Alexander and G Powell - The Simulation of Infiltration Rates and Air Movement in a Naturally Ventilated Industrial Building, *Proc 12th Air Infiltration and Ventilation Centre Conf,* Ottawa (1991) 273-284 ISBN 0-946075-53-3
5 A Davies, PhD thesis in preparation.
6 D Alexander, P J Jones, and H Jenkins, Tracking Air Movement in Rooms, *15th AIVC,* (1994) 448-492.
7 P J Jones and G E Whittle - Gateway II: Airflow Modelling, Dept of Energy, London (1992) pp11 ETSU S 1323.
8 P Jones and R Waters - The Practical Application of Indoor Airflow Modelling, *Modelling of Indoor Air Quality Exposure,* Niren L. Nagda (ed), 173-181, ASTM STP 1205,(1993) ISBN 0-8031-1875.
9 P J Jones, D Alexander and H Jenkins - Investigating the Effects of Wind on Natural Ventilation Design of Commercial Buildings, *EPSRC Project* (GR/K19129).

CHAPTER NINE

Specifying Environmental Conditions for Naturally Ventilated Buildings – A Consultants View

Chris Twinn BSc(Hons), CEng, MInst, MCIBSE
Welsh School of Architecture, UWCC, Cardiff, UK

Building environmental design

Engineering naturally ventilated buildings requires a very specific design service dealing with building form and fabric. Unlike building services it major input is at feasibility and conceptual stages of design. This design service provides a fabric performance specification normally delivered as a design report. At the strategic level this design service normally includes:

- Comfort
- Thermal environment
- Lighting
- Acoustics
- Energy consumption

Commonly a detailed analysis of comfort and the thermal environment is also provided as this is the area where most significant design influences are found. However it is increasing involving detailed analysis of the other aspects above.

Comfort tends to be considered under a separate heading from heating and cooling because it involves a wider range of psychological and well as physical factors. Thus it tends to consider a fuller range of comfort issues like perception of control and fabric radiant temperatures, unlike conventional temperature control which tends to relate to simply air temperature as the input to mechanical services design.

Given that natural ventilation has a cooling capacity of perhaps up to 50 W/m^2, a large proportion of effort tends to go into reducing solar gain from the more typical 100 W/m^2 for a perimeter room down to a more manageable 25 W/m^2.

Naturally Ventilated Buildings: Buildings for the senses, the economy and society. Edited by D. Clements-Croome. Published in 1997 by E & FN Spon. ISBN 0 419 21520 4

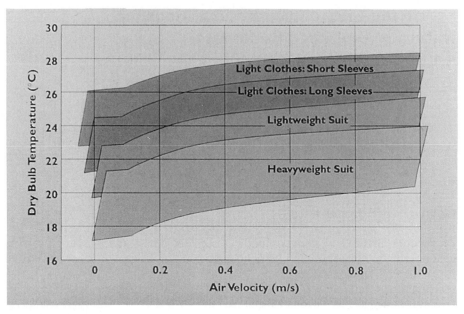

Figure 1   Diagram showing how a person's air temperature satisfaction range varies depending on their clothing and air movement.

Client's benefits

So what are the benefits of Building Environmental Design for the client?

The resulting buildings tend to have more flexible built into them. Effectively they tend to provide most of the internal environmental control as a function of the Shell & Core state. The subsequent fitout normally has little mechanical content. Yet they should be designed to have the flexibility of an in-built complete upgrade strategy, so should a tenant or occupier need it the space can have added, in part or in whole, mechanical ventilation, sensible cooling or full air conditioning. At lease end the tenant fitout can be stripped out to reveal a fully functioning base form with a minimum of extra work.

The buildings are simpler and potentially cheaper, so more money is available for increased floor area, the architecture, or alternatively, back into the clients pocket.

The design provides a predictable calibrated building that has a specified perfor-mance along the same lines as would be the case with an air conditioned building.

The buildings tend to be more robust in operation. Being simpler there are few items that can go wrong. In addition nature tends to be remarkable self limiting. As a simple example, as ventilation air gets warmer, so the buoyancy effect increases. Likewise the use of exposed thermal mass provides an inherent overload capacity.

Costs tend to be lower in use, be it for maintenance, energy, refurbishment, or life cycle costing.

The design has a more integrated form with potentially more 'elegant' solutions.

The specification

If one considers comfort, typically this would include a Predicted Mean Vote (PMV) level for perimeter rooms exceeded for a specified number of hours per year. This would relate to a dry resultant temperature. The choice of PMV level is dependent on a whole range of occupant psychological factors, including their ability to adapt, their expectations, and how much they feel in control.

The weather data basis is crucial as conditions are so weather dependant. The design will need testing separately against a number of worst weather conditions, including solar radiation peaks and air temperature peaks.

The specification will schedule out the full details of window form, positions, sizes, glass types, shading devices and opening areas.

If passive cooling techniques are to be included to reduce peak temperature condition the specification will describe how it is to be achieved. So if it involves nighttime ventilation cooling it will include the areas of thermal mass, its positions, density, conductivity, thickness, and finish. It will also include the method of ventilation control both for daytime and night.

Dependent on building form and the various natural ventilation routes the specification needs to record the physical requirements for atria, solar towers, thermal insulation. The zones for 'closer comfort' and 'looser comfort' conditions will need to be defined. Obviously, as with any design, the brief and design assumptions must be clearly recorded.

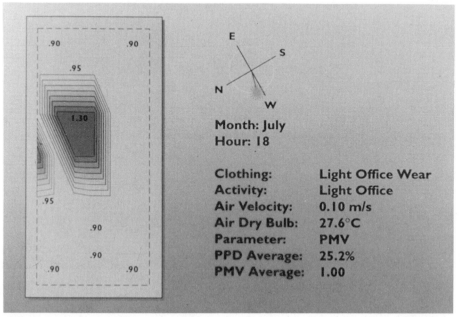

Figure 2 Thermal comfort computer analysis output showing a room floor plan with a shaded zone where direct sun radiation is penetrating the window shading to create a discomfort zone.

Project one: Anglia Polytechnic University – the project

Anglia Polytechnic University's (APU) new Learning Resource Centre in Chelmsford completed in September 1994, provides a 6,000 m^2 floor area to combine the functions of library, study areas, a TV studio and recreational facilities all under one roof. The brief established between the client and the design team was for a building that satisfied:

- a very tight capital cost budget
- a short construction period
- robust materials and finishes
- minimum maintenance
- low running costs
- a minimum of mechanical systems
- good day lighting
- can be fully used all the year round
- the ability to accept a computer on every desk.

The design process

The design process commenced with a week long workshop at which architects ECD and environmental engineers Ove Arup & Partners worked closely together to

Figure 3   South west elevation of Anglia Polytechnic University Learning Resource Centre. Note the windows are divided into lower vision and upper daylighting parts with trickle ventilation wall openings below.

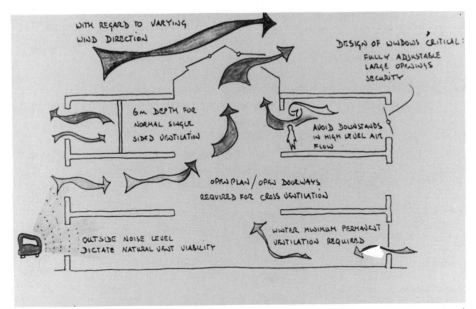

Figure 4 Concept diagram for ventilation design parameters. Similar diagrams considered cooling, daylighting and heating.

understand and discuss with the client representatives their needs and ways of fulfil them. This was based on the holistic approach necessary to achieve an integrated solution that bridges traditional professional boundaries. A range of alternative solutions were explored before narrowing down to the preferred option. During the week many issues were explored, even to the extent of daylight testing under the Bartlet artificial sky. There was also a meeting with the local planners with its influence on external aesthetic treatment and massing. By the end of the week perhaps 80% of building's fundamental design had been established.

One of the key features of this process is the dependence on the individual experience of the team members. At the conceptual stage of design there is little time for detailed engineering analysis of the many ideas discussed in the free flowing discussions. All engineering options had to be fully explained to the team before the building form can takes shape. It was at this stage that the principle of natural ventilation was established and the building fabric components necessary for passive cooling were agreed.

Key design objectives included:

A fully integrated facade as the primary climate modifier between inside and out providing:

- good views out,
- opening windows
- draught free background trickle ventilation for when windows are shut during cold weather

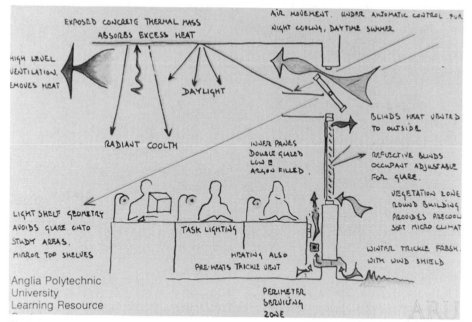

Figure 5 Concept diagram, developed during design workshop, of facade and perimeter study areas.

- very high levels of solar shading (blind inside triple glazing, ventilated to outside)
- daylight reflected deep into the building even when solar shading is in operation
- attenuation of excess perimeter daylight
- individual glare control
- high resistance to heat loss
- perimeter heating system
- the main services distribution
- good window security in all public areas.

Atria to provide:

- light and airy central circulation spaces
- a shape that widened at the top to maximise daylight and beneficial solar heat gain into the middle of a 30 m deep building
- reflective solar shading down atria sides to reflect light down toward the lower floors
- a stack assisted ventilation route up and out of the middle of a the deep plan
- a roof profile intended to assist ventilation exhaust by maximising leeward negative wind pressure coefficients

Internal finishes to include:

- large areas of exposed high thermal capacity surfaces that are inherently robust and need low maintenance

- light colours on selected surfaces to enhance daylight reflection
- carpets to act as sound absorbing surfaces

Building general arrangement

APU is located on a redundant industrial site between a busy road, an flour mill, student residential buildings and the river Chelmer. It is rectangular in plan, increasing in elevation at the points where the two central atria occur. The four storey elevation presents a formal face overlooking the road while the recreational facilities have a softer treatment onto the more attractive riverside.

The structure consists of an insitu reinforced concrete frame with screed topped floor slabs. The underside of the roof has precast concrete panels supported between an exposed steel frame. The underside of the floor slabs and the roof precast units are fully exposed to the rooms below with just a paint finish to permit them to be the main passive cooling surfaces.

Ventilation strategy

A detailed review of coincidental weather conditions in relation to the APU building form identified a number of peak design conditions of which the high air temperature and low wind combination was critical. For a significant proportion of the year a hybrid of part cross ventilation with stack influences is likely to dominate, but the sizing of building component (eg window sizes) is dictated by the pure stack critical case. However throughout, the design needed to be checked against the other combinations, for example, cold high winds.

The detailed analysis to size the components was an iterative process of balancing heat gain and hence air buoyance driving force against all the flow resistances the air passes though on route through the building. The objective is to obtain that balance at an appropriate room temperature to satisfy occupant's expectations. The air flow routes are many and form a network that also needs to be aerodynamically balanced.

The lowest floor has the greatest stack height driving force and so needs the most air flow throttling. The top floor has the least stack and so needs the least resistance to air flow. This balancing is achieved by varying the areas of opening windows on each floor, so the bottom floor has least and the top floor most, in this case by a factor of more than five. Getting adequate height separation between the top floor opening and the atrium vents, as well as adequate atrium vent area is also critical. Otherwise the vitiated air from the lower floors will find it easier to exit via the top floor windows so engulfing the occupants.

Security concerns precluded manual control of many low-level windows, so an arrangement of actuator driven clerestory windows are used in the public open plan areas. These are arranged in facade orientated groups on each floor, each group controlled by its own local room temperature sensor. These same sensors also deal with excessive wind effects simply by responding to the increased cold air infiltration. In cellular rooms where the staff occupants are likely to want to control their own conditions manual opening low-level windows are provided

During winter, experience has shown that windows are simply kept shut in naturally ventilated buildings. At times this can create stuffy polluted conditions. It is ironic that it is this that probably maintains adequate winter minimum indoor humidity levels.

At APU the balancing of these factors resulted in an engineered trickle ventilation arrangement. The local buoyancy effect generated by the perimeter heating inside its casing is used to draw air through fresh air vents from outside. The intention is to is obtain a continuous background tempered fresh air supply with a large replenished reservoir at the start of occupancy each morning. This avoids supplying large volumes of heated dry air to suit a theoretical maximum occupancy. Atria air sensors provide a ventilation override should air deteriorate. Central manual override ventilation switches are also provided.

Cooling strategy

For more than 90% of the year adequate cooling is available by using outside air via opening windows, assisted by the atrium stack. During periods with daytime peak outside air temperatures, cooler nighttime air is used to provide supplementary cooling. For design weather conditions and internal heat gains of about 43 W/m^2, the room dry resultant temperature is expected to exceed a peak of 27°C for about 45 hours per year, less than 2% of the building occupied period.

Warm room air temperatures are offset by using cool exposed high thermal capacity room surfaces to provide a lower radiant temperature source and, thereby cooler comfort conditions. For the radiant cooling maximum benefit, the high

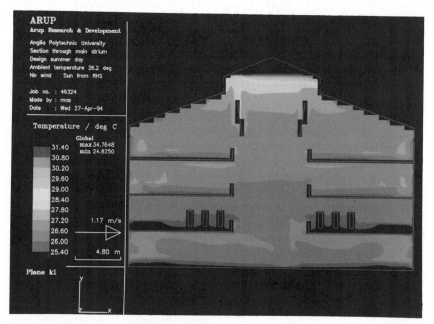

Figure 6 Computational fluid dynamics (CFD) computer modelling of natural ventilation air temperatures across the building section. Note the temporary temperature inversions and the influence of bookstacks on the first floor.

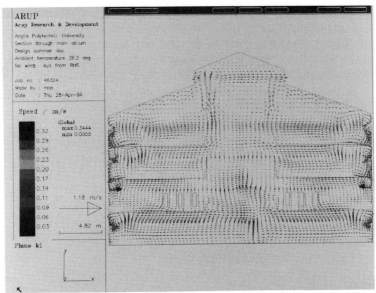

Figure 7 Computational fluid dynamics (CFD) computer modelling of air flow for natural ventilation in stack mode. This represents a time snapshot as natural ventilation has a continuously changing pulsing movement.

capacity surfaces must in practice be directly visible to the occupant. Nighttime ventilation is used to removed the heat absorbed by the surfaces during the previous day so the surfaces are available to provide cooling the following day. The atrium stack provides the ideal method for driving this nighttime ventilation.

For night cooling one of the key issues is the duration of ventilation needed to adequately cool the structure but avoid overcooling and a heating need the following morning. Direct measurement of surface temperature was discounted as it varies with position and surface depth. Likewise, monitoring of night exhaust temperature gives little clue, air entering at say 16°C can easily leave at 18°C irrespective of building fabric temperature, the limiting factors being the finite heat transfer are and contract time.

The approach taken at APU is to monitor the daytime room temperature degree-hours over a room set point and then terminate cooling when the night degree-hours is equals to it. The nighttime set point is in effect a theoretical average building fabric temperature which is continuously adjusted by a BMS self learning routine based on achieving comfortable conditions at the start of occupancy.

Considerable computer analysis was involved in developing the engineering designs. The software included a zonal air flow model and a thermal model. The latter considered room air temperature stratification, the building fabric as a resistor/capacitor representation, separate shortwave and longwave heat transfer using form factors, as well as algorithms for modelling ventilated cavity blind/glazing systems and for the light shelf heat redistribution.

An important design objective was to keep the controls simple with the minimum of overlapping control loops, operating modes and variables. There are many external and internal influences that if the BMS was to give then equal weight would have resulted in considerable complexity. Many of these influences are in practice secondary issues that need not warrant the complexity of their inclusion. Fundamentally the building has one daytime operating mode driven by the local room air temperature sensors. At times of sustained peak room temperatures this is supplemented by the night ventilation algorithm. Where practical room control has been delegated to the occupant. This includes the task lighting, window blinds and perimeter heating as well as windows in cellular staff areas.

Energy use

APU has many energy saving aspects summarized by:

- general avoidance of mechanical ventilation
- passive cooling techniques in lieu of mechanical cooling
- modest winter fresh air heating need.
- enhanced day lighting levels
- modest background artificial lighting levels
- daylight control of background lighting
- task lighting
- condensing gas fired boilers
- very low pump circulation heads
- domestic hot water preheat from bar cellar cooler heat recovery
- displacement ventilation and high efficiency heat recovery (and evaporative cooling recovery) for catering area mechanical ventilation

The building services delivered energy consumption target has been set at a low 82 kWh/m². A fairly conservative view has been taken for the target because the final figure will be greatly influenced by the way the building is used. The design placed more emphasis on reducing primary energy as this not only yields lower carbon dioxide emission targets, namely 40 kg/m², but also directly reduces the overall purchased energy costs.

Overall impressions

The procurement method chosen by the client was Design & Build with novation of his design team after tender stage. On the whole this allowed the client to achieve all his main objectives with a good value building.

There are various lesson that could be learnt, in particular with regard to commissioning. Part of this seems to be related to the distancing of the design team from the client by the novation process. The final commissioning and client acceptance personnel had limited understanding of the design principles. As a result items like the lighting control system and BMS settings provided as standard by suppliers did not reflect the building design intent or how the client eventually decided to use the building. However this may just be a reflection of the industry as a whole.

The APU building environmental approach is one of loose fit with the ability to accept local upgrade when or if necessary. The use of the top floor was changed immediately before occupancy with minimal adaption to the environmental systems.

The use of the computer room and with it the services provision has also been subsequently upgraded. The approach is one that acknowledges a building will undergo changes and that a small desk fan may be an appropriate occasional solution in terms of capital cost, running cost and environmental impact.

Through adherence to an environmentally aware approach by the design team and the client, the APU building has attracted an EC Thermie demonstration project grant. This will allow extensive in-use monitoring that should produce the feedback that the industry is so poor at gathering and using.

PROJECT TWO: Inland Revenue Building Nottingham – the project

The Inland Revenue Building (IRC) comprises 40,000 m^2 office space and its associated ancillary facilities.

It was completed in October 1994. An enlightened client brief, prompted by the intervention of the local residents, called for an environmentally friendly and naturally ventilated building. With this in mind, the architect, Michael Hopkins & Partners and the Engineers, Ove Arup & Partners, set about designing an innovative

Figure 8 Nottingham Inland Revenue centre showing glass staircase ventilation towers each with motorised raise/lower fabric top.

buildings using the strategy of maximizing daylight and providing engineered natural ventilation.

The design uses the building's structure, form and internal environmental needs to inform its architectural expression in a creative way. The success of this approach is reflected in the fact that the building has achieved the maximum possible points in a BREEAM assessment for new office buildings.

Form and construction

The site was previously industrial and is bordered by a main railway line, a flyover and a canal, with views towards the centre of Nottingham with its castle. A principle architectural objective was to extend the city grain across the canal onto the site. The scheme introduces seven free-standing courtyard and L-shaped buildings each with dominant corner staircase towers organised in a series of radiating bands of streets, buildings and gardens along a tree lined boulevard which extends across the site in a gentle curve.

The building uses cost-effective and fast track construction techniques to meet the programme and budget. These included the decision to prefabricate off-site both the solid brick piers and the precast wave form concrete floor slabs. The precast units span 13.6 m between the perimeter columns and provide exposed vaulted ceilings for all the lower floors. The top floor of each building overhangs with exposed roof steel trusses spanning 15 m.

Ventilation strategy

Fresh air enters the building through occupant controlled tilt and slide windows, passes through the office space and full height corridors to exit the building via the solar assisted corner ventilation towers. This route is kept open by using emergency released electromagnets on all the fire doors between the offices and the towers.

The fabric roof 'umbrella' of each ventilation tower can be raised and lowered to control the rate at which air is exhausted. The external walls of the ventilation towers are constructed of glass blocks. This allows the towers to absorb solar energy and so supplement the convective air current rising up the tower. Thus there is a direct link between the degree of solar radiation and the induced ventilation rate.

The ventilation design was tested using a number of modelling techniques. These included computer models and physical testing using the saline fluid method developed at the University of Cambridge. These confirmed that to be effective the ventilation towers needed a height of at least 7 m higher that the top floor they served. Accordingly the IRC top floors are ventilated separately with their own ridge vents and additional storey height.

The ventilation strategy leads to preferred rules for office layouts:

- generally open plan
- cellular offices on a 3.2 m planning grid with a minimum 1.6 m corridor for air to reach the towers
- solid partitions at right angles to the perimeter walls.

- fit-out screens should stop 200 mm below the lowest point of the wave form ceiling.
- 2.4 m high doors to cellular rooms which when open permit air passage at high level
- ventilation becomes single sided if cellular offices are fully enclosed and used with doors closed

Building environmental approach

The design incorporates many features (both passive and active) that help to maintain moderate internal conditions. Dominant among these are those that address the minimising of internal heat gains (see figures)

- Generous areas of triple glazing with mid pane blinds
- Glass light shelves that act predominately as shading to the perimeter office zones but not at the expense of day lighting levels deeper into the room
- Blinds in clerestory set at the optimum angle to allow in reflected light but preclude direct solar radiation
- High efficiency lighting complete with daylight dimming, to minimise lighting heat gain particularly at times of peak solar gain
- Projecting external brick piers provide solar shading.
- High thermal capacity exposed concrete soffit acts as heat sink absorbing excess daytime heat gains.
- High ceilings and ventilation towers as part of a defined ventilation path through the building.
- Occupant controlled below floor perimeter small fans allow windows to be closed in winter or on noisy east and south site boundaries.
- Perimeter fans used to purge the building of heat at night (up to 6 air changes per hour)
- Vaulted ceiling profile developed to avoid acoustic 'hot spots' and to have acoustic focus points below the carpeted floor level.
- Building Management System (BMS) sets default control settings for heating, ventilation and lighting.
- Occupant room override control of lighting, ventilation, blinds and heating.

Conclusion

The Inland Revenue offices promises to be an effective example of a truly integrated design. The dictates of an environmentally conscious brief have inspired the designers to use the building's envelope, structure and form as climate modifiers. This should result in a satisfactory internal environment without recourse to air conditioning. A distinctive architectural vocabulary has been generated by energy and environmental concerns, which also integrates an acknowledgement of the surrounding architectural context of Victorian warehouses.

CHAPTER TEN

Natural Ventilation in a Large Mixed Mode Building

David Arnold
Troup, Bywater & Anders, London, UK

Abstract

A large credit company has decided to relocate and centralise its offices and operations from a number of city centre sites in Northampton to the outskirts on a 'green' field site. Their present buildings are air conditioned but, the concept for the new building is an *'environmentally friendly'*, natural ventilated, low rise, one with a street style atrium. Only a small area of the building requires air conditioning for operational reasons. At an early stage of the conceptual design the client expressed certain concerns over whether natural ventilation would be adequate to maintain comfort in view of the extent to which the company relies on computers and associated peripheral equipment. The heat released from equipment in typical areas of their existing offices was surveyed and used as parameters in thermal models. Initial temperature predictions indicated that on the basis of natural ventilation alone internal temperatures would only exceed 26°C in the afternoons of a few days in a typical year. However, the client decided that the internal temperature should be limited to a maximum of 25°C during the warmest days of a typical summer. Intentionally providing a margin for exceptional summers and the risk of climate change. The client also required the option to install full air conditioning at some future time for investment purposes. Comparisons were made between full air conditioning, various mixed mode strategies. The design solution is a mixed mode building comprising both natural and mechanical systems that will operate predominantly in the natural mode. Natural ventilation is provided by fixed open trickle ventilators, opening windows, and automatically controlled vertical open lights in the atrium roof and side wall glazing. The mechanical systems include forced ventilation (displacement) and direct cooling (chilled beams), only intended for use when natural systems are inadequate. The paper is a design case study that describes the philosophy behind the design and, the analyses and modelling techniques used to determine the provision for natural ventilation.
Keywords: Air conditioning, Energy efficiency, Mixed mode, Natural ventilation, Opening windows, Overnight cooling, Passive cooling, Trickle vents.

Naturally Ventilated Buildings: Buildings for the senses, the economy and society. Edited by D. Clements-Croome. Published in 1997 by E & FN Spon. ISBN 0 419 21520 4

1. Introduction

A large credit card company has decided to relocate and centralise its offices and operations from a number of city centre sites in Northampton to the outskirts of the city on a green-field site. Their present buildings are air conditioned but, the concept for the new building is an *environmentally friendly* one, low rise and naturally ventilated, . Only a small area of the building requires air conditioning for operational reasons.

The new building is large, 37,000m² and, when taking into account the space required for car parking, occupies a compact site. Initial studies, by the Architect, considered three alternative plan forms, shown in figure1. The doughnut and prism plan forms would probably have provided better natural ventilation but the street form was chosen on the basis of organisational requirements and, in addition, had the investment advantage that it could be valued as three separate buildings. The complete building is, in effect, six buildings of between four to six stories, parallel to one another, either side of a street. An early design decision was to cover the street with a glazed roof and form an atrium that acts as an amenity and circulation space as well as a buffer zone.

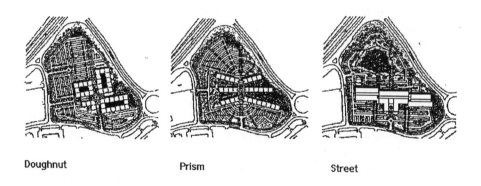

Doughnut   Prism   Street

Figure 1. Conceptual Plan Form Alternatives

Once the conceptual plan form had been decided analyses were carried on the building form and fabric to minimise energy use. The early design decisions included, a) placing the lower block south of the higher to provide maximum sunlight penetration and day lighting to the street, b) brise-soleils to the south facing windows to prevent excessive direct beam sunlight entering the offices in summer and, c) exposing the ceiling slab in the offices to limit the swing in temperature during the day. Subsequently it was decide to provide awnings to the street windows for local solar protection rather than treat the glass roof . In addition a large balancing pond was required to deal with storm water and it was decided to use this both for sprinkler water storage and, as a source for cooling and heat rejection. A cross section of the architect's design concept is shown in figure 2.

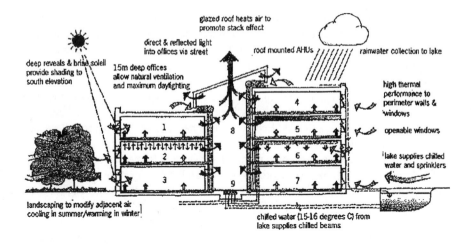

Figure 2. Architect's Design Concept

At an early stage of the design the client expressed concern over whether natural ventilation alone would be adequate to maintain thermal comfort, during hot weather, particularly in view of the extent to which the company relies on computers and associated IT equipment. Consequently the heat released from equipment, in typical areas of their existing offices, was measured and used as a parameter in thermal modelling. Initial temperature predictions indicated that, on the basis of natural ventilation alone, internal temperatures would only exceed 26°C in the afternoons of a few days in a typical summer. However, the client decided that the internal temperature should be limited to a maximum of 25°C during the warmest days of a typical summer, intentionally providing a margin for exceptional summers and the risk of climate change. The client also required the option to install full air conditioning at some future time, for investment purposes.

2. Options for environmental control

The options for heating, ventilating and cooling large buildings such as offices have, over time, been polarised into two separate and usually isolated strategies. Either, full air conditioning or, natural ventilation coupled with winter heating. However, research on both types of building, in the UK (1), has shown that occupants seem to be of the opinion that there is little to choose between them. A good naturally ventilated building can be almost as comfortable as a good air conditioned one and vice versa. Hawkes (2) has described this polarisation into either fully air conditioned buildings or those where natural systems predominate, as different approaches to the same environmental problem ie, determining the relative functions of building form, fabric and engineering. In the former the priority is to isolate the internal environment from the external using mechanical systems. In the latter the building envelope is given priority and configured to a) maximise the use of ambient energy and, b) achieve an effective balance between the use of the advanced automatic control of passive devices and the opportunity for the users of buildings to exercise direct control of their environment. Hawkes defines these strategies as exclusive and selective modes and characterises them as follows:

Exclusive Mode. Building orientation is unimportant and shape is compact to minimise interactions between exterior and interior. The internal environment is   predominately artificial and automatically controlled . Energy use is at an   almost   constant   rate irrespective of season and from generated sources.

Selective Mode. Building orientation is crucial and shape is formed to make   the best use of ambient energy. The internal environment is controlled by a mixture   of   automatic and manual means. Southerly windows predominate and are provided with solar controls to avoid summer overheating. Energy is from   ambient and generated sources peaking in winter.

Hawkes illustrates the difference between these approaches by reference to Olgyay's (3) principle of selective design, shown, adapted, in figure1. The principle is to make effective use of the natural environment and reduce the loads for mechanical systems from, for example with   air conditioned buildings, the gap between curves 1 and 5   to the gap between 3 and 5.

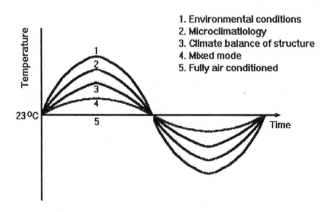

1. Environmental conditions
2. Microclimatiology
3. Climate balance of structure
4. Mixed mode
5. Fully air conditioned

Figure 3, Olgyay's principle of selective design with the addition of Mixed Mode

To meet the Client's brief this building needs to be almost chameleon like and change from the selective mode to the exclusive during extremes of weather. The basic philosophy being that mechanical systems are only used when natural means are inadequate. If now the requirement for constant internal conditions is relaxed to that of a free running building the gap can be further reduced to that between 3 and 4. The intention being to provide year round comfort but avoid the cost, energy penalty and consequential environmental effects of full year round air conditioning. The challenge of the client's brief is to provide a solution that allows the building to alternate between the two strategies and retain their respective advantages.

This technique of maintaining a satisfactory internal environment by alternating between passive and active modes has been used on a number of and has recently been given the title of *Mixed Mode* (1). It can be described as providing a comfortable internal environment using both natural ventilation and mechanical systems, but using different features of the systems at different times of the day or season of the year. In principle the mechanical systems include heating, mechanical ventilation and, or, mechanical cooling ie, refrigeration.

There are a wide range of interpretations of the term mixed mode but it has been in use, in one form or another, for around 20 years in the UK(4). A recent study of such buildings (5) concluded that they are generally successful, popular with the occupants and, able to convey the feeling of a prestige building. The approach takes advantage of the adaptability of human beings to accept and feel comfortable over a range of temperature during the course of a single day. In principle, the occupied spaces start the day as comfortably cool as acceptable and finish as comfortably warm as acceptable.

A simpler solution might seem to be to provide a conventional air conditioned building with openable windows but only use the air conditioning when internal temperatures indicate that it is required. This has been done in the past and the result is that with near constant internal temperatures there is no stimulus to open windows (6) and, use natural ventilation. The potential energy savings are lost and the building owner finishes up paying twice ie, for both openable windows and air conditioning. This lack of stimulus was overcome in one German office building by using different coloured lights that indicated to occupants whether they should open either, windows or, the air conditioning dampers (7). Although this solution may seem impractical in our culture, it more than halved the running time of the air conditioning. There is nothing new in the principle; some of the earliest air conditioned buildings in the US gave occupants the option to open either, the windows or, the use the air conditioning by opening dampers (8). Even installing an extract fan or a packaged air conditioner in a naturally ventilated room meets the definition however, the objective in this design is to optimise the systematic application of the principle to the entire building.

The main attractions of this method of providing cooling, to buildings that cannot be satisfied by natural ventilation alone, are two fold. Firstly, the introduction of openable windows into what might otherwise be a sealed building provides the opportunity for users of the building to exercise direct control over their local environment, both in terms of the rate of ventilation and, when outside temperatures are suitable, cooling. Secondly, by only using mechanical systems such as forced ventilation and/or cooling when necessary much less energy is used. In addition to these attractions most mixed mode systems leave open the option to change to conventional full air conditioning.

The energy use can be further reduced by taking advantage of cooler temperatures overnight and the thermal storage capacity of the building(9). This includes partitions, the floor slab, the inner portion of the outside wall and even furniture . Heat absorbed during the day, by the structure and fabric, is rejected by ventilating the building overnight. The technique is nothing new and is indeed a common feature of many traditional buildings and vernacular architecture(10). In this application it extends the times of the day and period of the year when mechanical systems can be avoided or, their use minimised. The technique also reduces the size of mechanical plant necessary to meet the peak cooling demands on the hottest days of the year. These advantages can only be attained if the building is adequately cooled, overnight, to reject heat absorbed during the day. This can be achieved on most of the nights of warm and hot days energy efficiently using natural ventilation. However, on the few nights of the year when the temperature does not fall enough mechanical systems can be used to provide the same effect.

The design team reviewed other options between natural ventilation and full air conditioning and chose mixed mode as the most suitable. They recognised that to justify the case for mixed mode, the comfort criteria must be met and, the overall life cycle cost should be less than with air conditioning, including the total cost of mechanical and natural systems. To this end the approach to the design of this mixed mode building was carried out in three stages:

1. The first was a conceptual design of the mixed mode building and systems.
2. The second was an intuitive analysis of the building supplemented by   preliminary calculations intended to, a) optimise the use of passive devices such as solar shading, natural ventilation and exposed thermal mass and, b) provide a basis for realistic cost estimates for the building fabric, services,   energy and maintenance.
3. The third stage, detailed design involved, a) calculations and thermal   modelling to refine the intuitive design, b) predictions of the internal conditions   resulting from the use of passive devices and, c)the duty of mechanical systems necessary to supplement the natural systems and maintain space   conditions within the design criteria.

3. Mixed mode concept

The principle of mixed mode is that the operating mode changes with the seasons, and within individual days, such that at any point in time the current mode reflects the external environment (weather) and takes maximum advantage of ambient conditions. Most of the time occupants are expected to modify their own local environment by using blinds, opening and closing windows, switching lights etc. as a reactions to normal stimuli such as glare, temperature, odour, stuffiness and illumination levels. Mechanical systems are only intended for use when natural systems are inadequate or in specific areas of the building for operational reasons.

An important feature of mixed mode is the need to allow the temperature to rise, within limits, during the working day. This is required to both stimulate the use of windows ie, provide natural ventilation and reduce the period of time when mechanical systems operate and, consequently the amount of energy used. This is in contrast to the basic philosophy of air conditioning that attempts to maintain a near constant condition of around 23°C, resulting in continual use of fans and pumps etc. irrespective of whether any cooling is necessary. The energy cost of fans and pumps in air conditioned buildings often represents around 60%, of the total required for air conditioning(10), and is one of the main energy savings of the mixed mode philosophy.

There is a lack of research on how people use local controls, such as windows, to modify their own local environment in buildings. This has a greater significance in naturally ventilated buildings where the internal environment depends on the building user reacting to some stimulus such as temperature, draught or, glare to change something, for example, open or close a window or operate a blind. This is a critical issue in the design of mixed mode buildings as three major questions need to be resolved:

1. How will people use opening windows in a naturally ventilated, mixed mode, building?

2. How do you provide the facility to open windows and avoid wasting mechanical cooling energy on hot days?
3. Taking account of the availability of mechanical systems, how large should the window openings be to provide adequate, cost effective, natural ventilation, in a mixed mode building ?

Research by Warren and Parkins (12) in the early 80s suggested that there are two distinct modes of window operation. In the first windows are used by occupants to provide a small area and this mode is largely independant of weather related variables. In the second, windows are used to provide a large open area and this is tied closely to outdoor temperature and sunshine. The explanation promulgated was that the first mode responds to indoor air quality and the second to indoor temperature. This corresponds with evidence from surveys of passively cooled naturally ventilated buildings in the summers of 1994 and 1995 (13). A number of buildings were visited and the proportion of windows open was typically in the range 5 to 10%. When the internal temperature rose above around 25°C the proportion of windows open rose significantly. The conclusion is that providing the 'trigger' temperature is not exceeded building users will respond in the first mode. In practice, providing that the internal space temperatures are not exceeded this hypothesis should limit window opening to about 10% of windows in a typical, naturally ventilated, building.

At the point when natural systems are no longer adequate to, restrict the rise or, fall, in temperature or, provide adequate ventilation, to within whatever limits have been set, mechanical systems are used to provide just sufficient heating, cooling or ventilation. An indication of when they are expected to operate related to the maximum daily external temperature is shown in fig. 4.

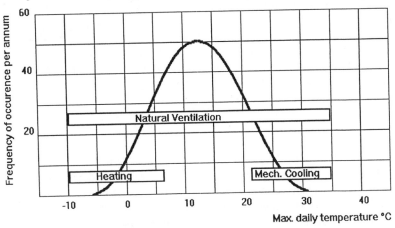

Figure 4. Typical operating modes related to maximum daily temperature

Three independent mechanical systems were chosen to work in conjunction with the passive systems, heating cooling and ventilation. Heating is required for obvious reasons, to warm the building and compensate for infiltration heat losses through trickle vents and building leakage. It is provided by simple hot water radiators beneath windows on both external and street facing walls. Each radiator has an individual thermostatic valve to provide local control for the building user.

Cooling is provided by either, chilled beams or, by cooling the air supplied by the supplementary mechanical ventilation system. Direct space cooling using chilled beams was selected because of the lower grade refrigeration necessary. The beams operate at a chilled water temperature of 15oC and are able to use pond water temperatures, for a large part of the year, without the need for mechanical refrigeration.

The supplementary mechanical ventilation system was considered necessary for several reasons, a)the depth of the offices between external windows is typically 40m and there was concern over achieving adequate ventilation across the open span on still days, b) similarly, during winter there was concern that building users close to external windows will keep them closed with the same result, c) hot days are often associated with cool nights and when these occur it could be advantageous to use the mechanical ventilation to increase the overnight cooling effect and reduce the refrigeration energy cost the next day and, d) should for any reason isolated areas be under ventilated by natural means the mechanical system could be used to compensate.

The design process must determine the capacity or extent of provision of the natural and mechanical systems. Part of this process is selection of the changeover point from passive to active systems, this will change depending on the extent of the provision of each sub-system. For example in the natural mode as the outside temperature increases more openings will be required to maintain the same internal condition. But, in this case a mechanical system is provided therefore, there is the opportunity to limit the provision of opening windows to a lower area than would be provided in an entirely passive building. However reducing the number of openings will incur an energy penalty. One means of approaching an optimum solution would be to carry out life cycle cost analyses but the final decision must take into account other subjective and practical issues such as, whether the proportion of windows 'feels right' for a naturally ventilated building, window cleaning and facade maintenance.

A similar choice arises in the provision of mechanical cooling. This can be provided by either the chilled beams or mechanical ventilation. Although both are required to meet the peak cooling situation, at other times there is an opportunity to use one or the other. It has been shown that the most energy efficient can be selected using the outside air temperature and the energy requirements of each system(14).

4. Detail design - natural modes

The size and distribution of glazed openings in the building, windows, doors, roof lights etc, were selected by the architect on the basis of providing adequate internal illumination, outside awareness and, of course, appearance of the facade. Most of these openings could be used for ventilation but not all of them are necessary to achieve satisfactory natural ventilation for a mixed mode building. Preliminary estimates using thermal modelling and other research (15) suggests that the optimum rate for overnight cooling is around 5 air changes per hour. When considering daytime natural ventilation the limits are probably imposed by: the comfort of occupants sitting close to (in charge of) the windows, the local temperature and the nuisance of papers lifting from desks or blowing around. The rate selected for overnight cooling of 5 air changes per hour would seem appropriate to meet these constraints. This equates to 3.75 l/s of outside air per square metre of 2.7m high office.

This stage of the design process included :

1. Confirming that the initial distribution of openings, based on the architect's concept, would adequately provide the selected rate of natural ventilation throughout   the building, including the 10% or so of cellular offices.
2. Selecting appropriate parameters for natural ventilation modelling , ie. wind speed and direction, internal and external temperatures, maximum and minimum size of openings, behavioural characteristics under manual window operation,   effect   of   automatic window operation.
3. Estimating air flow rates under range of conditions to compare with initial estimates and optimise the original selection.
4. Modelling thermal performance under the predicted airflow rates to ensure natural ventilation achieves design objectives.

The initial selection of ventilation openings was intended to provide controlled natural ventilation for most of the year. Whilst it was believed that the open plan areas could be predicted relatively easily there was greater concern about cellular offices. The actual locations of cellular offices are unknown as flexibility must be provided to allow them to be relocated. However, the client brief indicates that they will not exceed a maximum of 10% of the floor area and most but, not all, will be located against external walls. Only a few cellular offices will be located with windows opening into the atrium. Each of the typical cellular office locations was included in the modelling.

The intention in the concept design was to use the atrium roof lights in conjunction with the trickle vents and windows to control the rate of ventilation, both during the day and in overnight cooling mode. Consequently a decision was taken to limit the provision of automatically controlled vents to the atrium, in the roof and glazed end walls. Consideration was given to providing automatic control of the windows on the outside. This was rejected largely because of the uncertainty of the extent and location of future cellular offices adjacent these facades, which would negate their use and, on the grounds of appearance, cost and maintenance.

Each window will have a permanent trickle vent and opening lights. The size of the trickle vent, 160 cm² per 3m bay, was based on research by Perera et al(16) on the use of trickle vents for background ventilation. The architect's facade design allowed a range of possible opening windows and, as the facade modules are asymmetric relative to the partitioning modules, this resulted in three principal configurations with different openable areas. The sizes of which were modelled for comparison. The concept for the atrium has automatic opening vents in the roof and side walls that can be used in conjunction with the trickle vents and any open windows to control the rate of natural ventilation. The range of openings modelled in addition to the permanent trickle vents, and their intended use, is shown in Table 1.

Table 1.

Natural Ventilation Openings

Opening > Season v	Windows	Roof Vents	Atria Wall Vents
Winter Day	Manual	No	No
Winter Night	Closed	No	No
Cool Day	Manual	Auto	Auto
Cool Night	Closed	Auto	Auto
Warm Day	Manual	Auto	Auto
Warm Night	Man' if poss'	Auto	Auto
Hot Day	Manual	Auto	Auto
Hot Night	Man' if poss'	Auto	Auto

Initial predictions of the natural ventilation flow rates were carried out using manual techniques however, coupling the effects of all cells in the building proved practically impossible. And, as cost estimates indicated that the additional cost of making a single window light openable is around £200 to £300 per opening (1995 prices) and similar cost differences would apply to opening lights in the roof it was decided to use a computer model to examine the cost effectiveness of each component. The program (16) was then also used to predict most other cases of natural ventilation.

The model used was BREEZE, it is a multi-cell model that calculates air flows throughout a building by treating the building as a series of cells connected by air paths. It assumes that the inter cell air flows relate to a pressure difference by the common equation(16):

$$Q=AC_d(2(Dp)/r)^n$$

Where:

Q = volume flow rate of air through opening
A = area of opening
Cd = coefficient of discharge of opening
Dp = applied pressure difference
r   = density of air
n   = exponent relating volume flow to applied pressure difference

The driving pressures are calculated from the buoyancy and wind forces and an iterative procedure is used to calculate internal pressures and air flows that satisfy conservation of mass. The pressure coefficients were selected by BRE, from their library, who also undertook the modelling exercise.

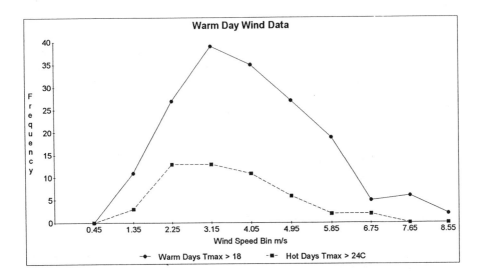

Figure 5. Wind Speed data on warm and hot days.

The clients brief intentionally allows margin, in terms of internal temperatures, for the eventuality of climate change. Therefore, to minimise risks that some local climate change has already occurred the most recent, most local, weather data available was used as a basis for the ventilation modelling. The best data available was from Bedford, just over 30km south east of the site. This data was only available for seven complete years from 1987 to 1993. An analysis of this data, figure 6, shows that the wind speed on warm days is similar to that on hot days but is significantly less than the design figure of 4.5 m/s given in the CIBSE Guide (16). The speed used for prediction purposes was 3.6 m/s (8 mph) for summer modelling and 7.2 m/s (16 mph) for winter with external temperatures of 18°C and 0°C respectively.

The results of the ventilation rate predictions were used as parameters for the dynamic thermal modelling. The simulation model was TAS and used to, a) predict the distribution of internal temperatures with and without mechanical cooling and b) to estimate using dynamic methods the minimum cooling necessary to meet the design brief.

5. Results

Several different cases were modelled to both predict the rate of natural ventilation under various modes of manual and automatic control and secondly to compare the impact on the rates of ventilation of the options of opening lights in the glazed openings. They included:

1. The impact of changing the proportion of opening lights in windows and roof glazing on ventilation rates in open and cellular offices.

2. The rates of ventilation at different percentages of windows open under manual operation.

3. The rates of ventilation available for overnight cooling using the automatically controlled openings in the atrium and trickle vents.

4. The rates of ventilation in the winter case with practically all openings closed except the trickle vents to assess heat losses.

5. The equivalent rate of fresh air ventilation achieved in the north offices with the prevailing wind from the south and the south roof lights open.

A selection of the results is included to illustrate the effects on design decisions.
Table 2. shows the results of the initial calculations to analyse the initial provisions of opening windows and roof lights. The exercise included the three configurations of opening window resulting from the facade design and partition lines. The roof lights are only open on one side of the atrium. and all openings are top hung and open to 15°.

The objects of this exercise were to a) determine the minimum area of window necessary to achieve the selected design ventilation rate with the windows open and, b) examine the effect of reducing the number of roof lights. The results are in air changes per hour for open plan and cellular offices, The cellular offices are in three locations, adjacent the, north and south outside walls and, the street atrium.

Table 2. Comparison of Window Opening Lights
(Ventilation rates in air changes per hour)

Opening Window m² per 3m	Roof lights - 50% opening				Roof lights - 33% opening			
	Open	North	South	Street	Open	North	South	Street
1.36	4.8	3.9	4.3	0.13	4.4	3.9	4.3	0.13
2.16	6.0	5.1	5.7	0.16	5.7	5.1	5.7	0.16
3.52	7.8	7.0	7.9	0.22	7.4	7.0	7.9	0.22

Table 2 shows that, apart from the cellular office adjacent the atrium, 2.16m² of opening window lights will be necessary per 3m bay of open plan office or cellular office to achieve the target of 5 air changes per hour. The reduction in openable roof light from 50 to 33% has a relatively small effect on the size of openable window necessary. This result of 2.16m² per bay was then used as a basis of the next exercise to determine the ventilation rates at various percentages of windows open. The results are shown in figure 6.

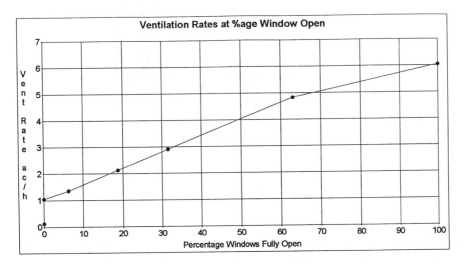

Figure 6. Relationship between the percentage of windows fully open and the predicted rate of whole building ventilation.
(Note The two values shown at 0% open show the effect of one side of the roof lights being open or closed.)

Figure 6 shows, by interpolation, that the natural ventilation rate at the anticipated window opening of 10% with internal temperatures below 25°C will be 1.6 air changes per hour.

The next exercise was to determine the range of automatic control available using the opening lights in the atrium roof and side walls to provide natural ventilation for overnight cooling.

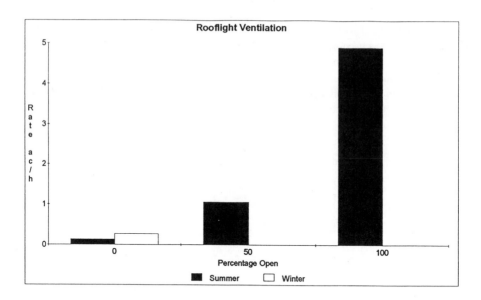

Figure 7. Ventilation using atrium automatic vents

Figure 7. shows that from 0.1 up to 5.0 air changes per hour can be achieved by control of the atrium vents. Whilst this shows the whole building air changes most of the circulation of air is within the atrium. With the windows closed, cooling of the offices spaces will be indirect and depend on convection generated between the warmer office and cooler atrium. The model does not take into account the affect of negative thermal buoyancy within the atrium. This is likely to enhance the effect of overnight cooling. Figure 7. also shows that the winter heat losses from the permanent trickle vents are not great. With all windows and roof lights closed the predicted ventilation rate including leakage is 0.13 air changes per hour.

Although the model will calculate the ventilation rate from any one of 12 wind directions, for simplicity, the cases above were all based on the prevailing wind direction, taken as 30° south of west. Under these conditions a significant proportion of the air ventilating the north offices will have passed through the south offices. Obviously the same thing happens in the opposite direction when the wind is from the north. It was of interest to know what the equivalent air change would have been if all the air entering the space was perfectly fresh ie, from outside. The BREEZE model determines this by simulating a uniformly distributed tracer gas transported around the building on the air flows. The results for offices at third floor level are shown in figure 8. as equivalent air change rates that take into account the *freshness* of the air.

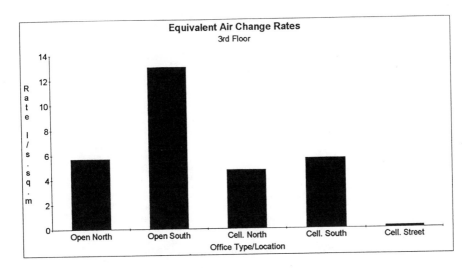

Figure 8. Equivalent air change rates

The predicted ventilation rates were used as input to a thermal model to predict the internal temperatures both with and without cooling. Figure 9. shows the frequency distribution of internal resultant temperatures using controlled natural ventilation overnight without mechanical cooling. The frequency of occurrence is hourly and each temperature represents the lower limit of a one degree band.

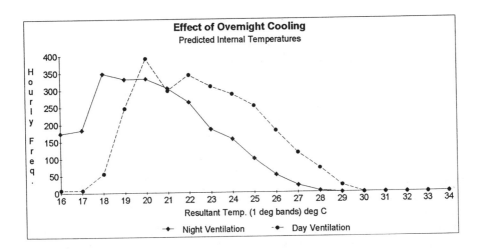

Figure 9. Temperature distribution using passive overnight cooling

The results confirm show that the temperature will equal or exceed 26°C for 79 hours in the course of a typical summer. The basis for these simulations was the records for Kew for 1967.

D Arnold

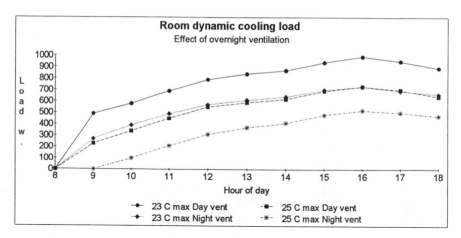

Figure 10. Peak room cooling load estimates using dynamic simulation.

The results in figure10. compare the cooling necessary to limit the maximum internal temperature to either 23oC or, 25oC with and without the use of overnight cooling. They are based on a near design day condition with maximum and minimum outside temperatures of 28.7oC and 15.2oC. They show the reductions in cooling loads that can be achieved by using overnight cooling and allowing the internal temperature to rise during the course of the day in hot weather.

5. Discussion

The comparison of opening vent sizes in table 2. shows clearly the difference in ventilation rates that can be attained between offices with external walls and those without. Reasonable rates of natural ventilation can achieved in all of the office spaces except the internal ones with windows onto the internal street. The reason is the relatively large unrestricted path for air flow in the adjacent open plan areas. It was decided to provide supplementary mechanical ventilation locally to any cellular offices in order to maintain natural ventilation to the majority of space. The results in table 2. also show that opening window configuration b), 2.16m^2, is necessary to meet the design natural ventilation rate of 5 air changes per hour, with the windows open, and this is appears to be largely unaffected by the size of roof vent. The reason for the relatively small influence on the roof vent size is the basis for comparison ie, all windows open. In practice the opening size and relation to the prevailing wind will have a significant affect on the ventilation rate.

The reason for modelling the effect on ventilation rate of various proportions of windows open was the risk that if too many are open during hot weather the affect of the mechanical cooling will be lost and the energy wasted. The results in figure 6. show that at the predicted opening proportion of 10% the natural ventilation rate will be 1.6 air changes per hour. This 'cooling load' is treated as an infiltration gain in the thermal model. There is similar concern in the winter condition that even when all windows are closed the trickle vents will provide too much ventilation and an excessive heat loss. The results, shown in figure 7, include building leakage and indicate a rate of 0.13 air changes per hour under summer design conditions and 0.27 for winter. This is only half of the value originally anticipated from the source research but is explained by the large depth of this building. This chart also show the range of ventilation rates that can be achieved, for overnight cooling and other ventilation, by the use of the automatic roof and permanent trickle vents. With all windows closed it is effectively from less than one quarter to five air changes per hour (based on 50% of the vertical roof light opening area). It is uncertain whether for security or other reasons if it will be possible to leave windows open overnight and, as this may change with time, the design is not based on this assumption.

The results from the thermal model confirmed the original estimates quite closely. In particular the distribution of internal temperatures, figure 9, reinforced the early advice to the client. The results show that with overnight cooling 26°C is exceeded for 79 working hours in the typical weather year used.

It is important that there is reasonable correlation between the design techniques used at each stage ie, intuition based on experience, simple models and the type of models ultimately used for these exercises. This allows solutions to be developed and be refined without the hiatus that might occur if the final detailed model gave radically different results.

The thermal model was also used to refine the estimates of room cooling loads under peak conditions. The influence and the opportunity for reducing refrigerating plant size on mixed mode peak limiting systems by comparison to conventional air conditioning is shown in figure 10. The combination of measures of, overnight cooling, exposed thermal mass and, allowing the temperature to rise during the day have reduced the peak load to almost one half.

6. Conclusions

The purpose of this paper was to demonstrate the design process for a relatively new type building, in terms of its environmental control, and show how the approach differs from the process for more conventional buildings of a similar scale. The goals were to, respond positively and imaginatively to the challenge of the design brief and, seek design solutions that could be audited. To this end the paper describes the design philosophy and methodology developed to respond to the challenge of the client's brief, an environmentally friendly building, naturally ventilated and, with cooling to limit the internal temperature to 25oC maximum.

The philosophy is to provide a mix of passive and active techniques that allows the building to maintain internal conditions in an energy efficient by alternating between natural and mechanical modes. This type of building has only recently been christened Mixed Mode but, the concept has been in use in one form or another for at least 20 years. The design solution has evolved from the designs of these earlier projects.

Key questions that had to be resolved as part of the design philosophy included a) how would people use the windows? and, b) would this result mechanical cooling, and energy, being lost through wide open windows in summer? This was resolved by the use of research into the behavioural characteristics of people in naturally ventilated buildings which indicates that there are two ways in which people use windows. The first is to obtain a small opening to provide air quality ie, dilute odours etc. The second is to create a large opening to provide a cooling effect. The move from the first to second mode being triggered by a rise in temperature. This research and visits to passively cooled buildings suggest that providing the *trigger* temperature is not exceeded building users are not stimulated to the second mode. Analysis of the research and experience of other naturally ventilated buildings indicate that the *trigger* temperature for most people is above the design peak temperature for this building of 25oC .

The natural ventilation system is intended to provide ventilation and cooling under local control in the offices throughout the year. During warm and hot periods it will be used overnight to pre-cool the building for the next day to reduce the need for refrigeration and energy. Each these modes of operation were modelled in detail and the results used to refine the size and allocation of natural vents. These result were then used as parameters for thermal modelling. This technique was used to confirm the preliminary estimates of the distribution of temperatures in the natural mode and to dynamically size the room loads for mechanical cooling. Use of the detailed models, a)exposed potential economies in the design of the natural vents and b) indicated that the combined measures of, overnight cooling, exposed thermal mass and, allowing the temperature to rise during the day reduced the peak cooling load necessary to almost one half of that required in a fully air conditioned building.

7. Acknowledgements

Barclaycard and Fleetway House Construction Ltd. for permission to use this project as a case study.

Fitzroy Robinson Ltd., Architects for permission to reproduce concept design.

Martin Smith of BRE for his contribution to the paper and undertaking the BREEZE analysis so enthusiastically.

8. References

1. Bordass, W.T et al 1995 " Comfort control and energy efficiency in offices" BRE Information paper IP 3/95.
2. Hawkes, D.1982 " Building shape and energy use", in "The Architecture of Energy," Longmans, London.
3. Olgyay, V 1963 " Design with Climate ", Princeton University Press, Princeton NJ.
4. Arnold, D. 1978 "Comfort air conditioning and the need for refrigeration" ASHRAE Trans. Vol.84, Part 2.
5. Willis, S. Fordham, M. and Bordass, W.1994 "An overview of technologies to minimise or avoid the need for air conditioning - Final report" CR 142/94 Building Research Establishment, England
6. Warburton, P. 1995 Private communication concerning the Royal Life Insurance building in Coventry, England.
7. Fitzner, K. 1993 " Air conditioning on request, an air conditioned building with windows that can be opened " Clima 2000 Conference London.
8. Worsham, H 1929 " The Milam Building ", Heating, Piping and Air Conditioning Vol.1, July 1929.

9. Fathy, H. 1986 "Natural energy and vernacular architecture", The University of Chicago Press

10. Energy Efficiency Office 1991 " Energy consumption guide - Energy efficiency in offices ", Department of the Environment, UK.

11. Arnold, D. 1994 "Integrating Thermal Storage in Buildings" RIBA-CIBSE Conference, Towards Zero Energy Buildings, London.

12. Warren, P.R. and Parkins, L.M. 1984 " Window opening behaviour in office buildings" ASHRAE Transactions 1984 Part 1. Paper AT 84 20.

13. Arnold, D. 1995 " Case studies of passively cooled naturally ventilated buildings" to be published

14. Tindale, A.W., Irving, S.J. and Concannon, P.J. 1995 "Simplified method for night cooling" CIBSE National Conference, Eastbourne, October 1995

15. Perera, M.D.A.E.S., Marshall, S.G. and Solomon, E.W. 1993 "Controlled background ventilation for large commercial buildings", Building Serv. Eng. Research and Technology, CIBSE, London.

16. Smith, M. 1991 "Breeze user manual 5.1" Building Research Establishment, Garston, UK.

17. CIBSE 1986 Guide Book A

Index